Things Fall Apart?

Studies in Environmental Anthropology and Ethnobiology

General Editor: **Roy Ellen**, FBA
Professor of Anthropology, University of Kent at Canterbury

Interest in environmental anthropology and ethnobiological knowledge has grown steadily in recent years, reflecting national and international concern about the environment and developing research priorities. 'Studies in Environmental Anthropology and Ethnobiology' is an international series based at the University of Kent at Canterbury. It is a vehicle for publishing up-to-date monographs and edited works on particular issues, themes, places or peoples which focus on the interrelationship between society, culture and the environment.

Volume 1
The Logic of Environmentalism: Anthropology, Ecology and Postcoloniality
Vassos Argyrou

Volume 2
Conversations on the Beach: Fishermen's Knowledge, Metaphor and Environmental Change in South India
Götz Hoeppe

Volume 3
Green Encounters: Shaping and Contesting Environmentalism in Rural Costa Rica
Luis A. Vivanco

Volume 4
Local Science vs Global Science: Approaches to Indigenous Knowledge in International Development
Edited by Paul Sillitoe

Volume 5
Sustainability and Communities of Place
Carl A. Maida

Volume 6
Modern Crises and Traditional Strategies: Local Ecological Knowledge in Island Southeast Asia
Edited by Roy Ellen

Volume 7
Traveling Cultures and Plants: The Ethnobiology and Ethnopharmacy of Human Migrations
Edited by Andrea Pieroni and Ina Vandebroek

Volume 8
Fishers and Scientists in Modern Turkey: The Management of Natural Resources, Knowledge and Identity on the Eastern Black Sea Coast
Ståle Knudsen

Volume 9
Landscape Ethnoecology: Concepts of Biotic and Physical Space
Leslie Main Johnson and Eugene S. Hunn

Volume 10
Landscape, Process and Power: Re-evaluating Traditional Environmental Knowledge
Edited by Serena Heckler

Volume 11
Mobility and Migration in Indigenous Amazonia: Contemporary Ethnoecological Perspectives
Edited by Miguel N. Alexiades

Volume 12
Unveiling the Whale: Discourses on Whales and Whaling
Arne Kalland

Volume 13
Virtualism, Governance and Practice: Vision and Execution in Environmental Conservation
Edited by James G. Carrier and Paige West

Volume 14
Ethnobotany in the New Europe: People, Health and Wild Plant Resources
Edited by Manuel Pardo-de-Santayana, Andrea Pieroni and Rajindra K. Puri

Volume 15
Urban Pollution: Cultural Meanings, Social Practices
Edited by Eveline Dürr and Rivke Jaffe

Volume 16
Weathering the World: Recovery in the Wake of the Tsunami in a Tamil Fishing Village
Frida Hastrup

Volume 17
Environmental Anthropology Engaging Ecotopia: Bioregionalism, Permaculture, and Ecovillages
Edited by Joshua Lockyer and James R. Veteto

Volume 18
Things Fall Apart? The Political Ecology of Forest Governance in Southern Nigeria
Pauline von Hellermann

Things Fall Apart?

The Political Ecology of Forest Governance in Southern Nigeria

Pauline von Hellermann

berghahn
NEW YORK • OXFORD
www.berghahnbooks.com

First published in 2013 by

Berghahn Books

www.berghahnbooks.com

© 2013 Pauline von Hellermann

All rights reserved. Except for the quotation of short passages for the purposes of criticism and review, no part of this book may be reproduced in any form or by any means, electronic or mechanical, including photocopying, recording, or any information storage and retrieval system now known or to be invented, without written permission of the publisher.

Library of Congress Cataloging-in-Publication Data
Von Hellermann, Pauline.
  Things fall apart?: the political ecology of forest governance in southern Nigeria / Pauline von Hellermann. -- First edition.
      pages cm -- (Environmental anthropology and ethnobiology; v. 18)
  Includes bibliographical references.
  ISBN 978-0-85745-989-3 (hardback: alk. paper) -- ISBN 978-0-85745-990-9 (institutional ebook)
  1. Forest management--Political aspects--Nigeria 2. Forest policy--Nigeria I. Title.
  SD242.N5V66 2013
  333.7609669--dc23

2013005540

British Library Cataloguing in Publication Data

A catalogue record for this book is available from the British Library

Printed in the United States on acid-free paper.

ISBN 978-0-85745-989-3 (hardback)
ISBN 978-0-85745-990-9 (institutional ebook)

For
Ohehen and Okechukwu

In memory of
Jonathan Ezele and little Constantin

Contents

List of Maps and Figures	ix
Acknowledgements	xi
Maps	xiii
Introduction	1
1. Ecology and Politics in the Benin Kingdom	20
2. Separating Farm and Forest: Reservation and Dereservation	45
3. Managing the Forests: Logging and Regeneration	86
4. Reinventing Farm and Forest: The Changing Forms of Taungya Farming	126
5. Okomu National Park: A Postscript on Conservation	145
Appendix. Administrative History of Edo State	162
Bibliography	163
Index	187

List of Maps and Figures

Maps

1. Map of Nigeria, showing Edo State and Okomu Reserve	xiii
2. Okomu Reserve in Edo State, Southern Nigeria	xiv
3. Okomu Reserve	xiv
4. Existing and new reserves demarcated under the Benin Forest Scheme in 1937	56

Figures

1. Statue of Aruaran in the centre of Udo	26
2. Path leading to Igueze village	37
3. The *inyator* of Igueze village	38
4. Odighi lake near Udo	40
5. A Yoruba cocoa farm in Okomu Reserve	76
6. A plantain farm in Okomu Reserve	77
7. Newly cleared land within Okomu Oil Palm Company	78
8. Oil palm at Okomu Oil Palm Company	79
9. Rubber trees at Osse River Rubber Estates Ltd.	79
10. Early logging in Benin forests	91
11. Silvicultural experiments at Sapoba Research Station	98
12. Typical log transport on an old Steyr vehicle	117
13. Recently cut *Ceiba* tree near Iguowan	122
14. A Taungya farm in Okomu Reserve	136
15. An overgrown old rubber plantation on Igueze community land	138
16. Path leading through Udo community land in Ogbe quarter	138
17. Landsat satellite image of the northern part of Okomu Reserve	139

Acknowledgements

I have received invaluable assistance and support from numerous organisations, colleagues, friends and family while undertaking this research and writing this book. I am grateful to the following institutions: The Economic and Social Research Council (ESRC) and the Natural Environment Research Council (NERC), who provided a joint student research grant; The British Academy, who provided me with a Postdoctoral Fellowship and a Small Research Grant; and the European Commission, who provided me with a Marie Curie Postdoctoral Fellowship as part of a larger research project on the Historical Ecology of East African Landscapes (HEEAL) at York University.

In Nigeria, permission to carry out the research was granted by the Nigeria National Parks Board. I would particularly like to thank Okomu National Park for their support during my research stay. I also had institutional support from the Nigerian Conservation Foundation (NCF) and the A.G. Leventis Foundation. I would like to thank the elders of Iguowan and the late *Uwangue N'Udo* for permitting me to conduct research in these places, and the managers of the Osse River Rubber Estates Ltd (ORREL), the Okomu Oil Palm Company (OOPC), and the Carmelite Monastery in Enugu for providing much logistic support and accommodation. The members of staff at the Nigerian National Archives in Ibadan and Enugu were very helpful, as were the staff of the Ministry of Agriculture and Natural Resources (MANR) in Benin City and Iguobazuwa and the Federal Surveys Department in Lagos, and many members of the History, Geography and Forestry Department at the University of Benin.

Above all, I am greatly indebted to a large number of individual people who helped to make my time in Nigeria so rewarding and enjoyable: Patrick Darling, Phil Hall, Georgina Okpiaifo, Adaku Nwanguma, Maria, who so kindly let me stay in her room in Iguowan, Philip Carty, Vincent and Sophie Verwilghen, Luluk and Harold Williams, Olivier and Patience Bonnefoy, Luca and Maria Teresa Tiboni, Rebecca Hanlin, Adebisi Sowumni and Saheed Aderinto. I would like to express my deepest gratitude to Alfred Ohehen, Egonu Okechukwu Boniface, the late Chief *Oliha N'Udo* Jonathan Ezele and Raphael, who have been wonderful guides and assistants, who taught me so much, and whose friendship I value dearly. Last but not least, I am indebted to all the people of Okomu, Udo and Benin City who so generously gave me their time for interviews and conversation and who let me participate in their lives – the openness with which I was received not only hugely benefitted my research but also made my time in Nigeria unforgettable. Back in the UK, I thank Richard Lowe, Peter Henry, David Ward and Donald McNeil for sharing their libraries and memories of their time in Nigerian forests with me.

I thank my doctoral supervisors John Peel, Jean Maley, Vinita Domodaran and above all James Fairhead, who greatly improved my thinking and writing. I benefited from the stimulating environments of SOAS, Sussex, York, and now Goldsmiths, where many conversations with faculty and students contributed to this work. In particular I thank Phil Stott, Sian Sullivan, Charles Gore, Richard Grove, Peter Lloyd, Grace Carswell, Elizabeth Harrison, Paul Lane, Kathryn Tomlinson, Nick Nisbett, Dinah Rajak, Rebecca Prentice, Dorte Thorsen, Sally Brooks and especially David Pratten, who has been an invaluable help throughout. I also thank William Beinart, David Anderson, Barbara Harriss-White, Alfred Grove and especially Bill Adams, who have also provided much support and inspiration. In addition, I am grateful for helpful comments on my writing and ideas from Uyilawa Usuanlele, Kojo Sebastian Amanor, James Fenske, William Clarence-Smith, Alan Grainger, Kalyanakrishnan Sivaramakrishnan, Roy MacLeod, Philip Alsworth-Jones, Paula Ben Amos, John Thornton, Alan Ryder, Arthur Mason, Bjorn Sletto, John Manton, Rhiannon Stephens and Neil Armstrong. In the latter stages of revision my manuscript was significantly improved by the insightful suggestions of my anonymous reviewers, and I am grateful for the guidance of Roy Ellen and Ann Przyzycki DeVita at Berghahn Press.

My family has been part of this project throughout its germination. During the final year of my doctorate I received generous help from my parents Manfred and Dorothee von Hellermann, which I thank them for. In recent years it has been my new family that has had to live with the production of this work: my children Clara and Otto (lying on my lap as I type this) and especially my husband, James Hampshire. I am deeply grateful for his constructive criticism and support, which have been crucial for bringing this work to completion.

Maps

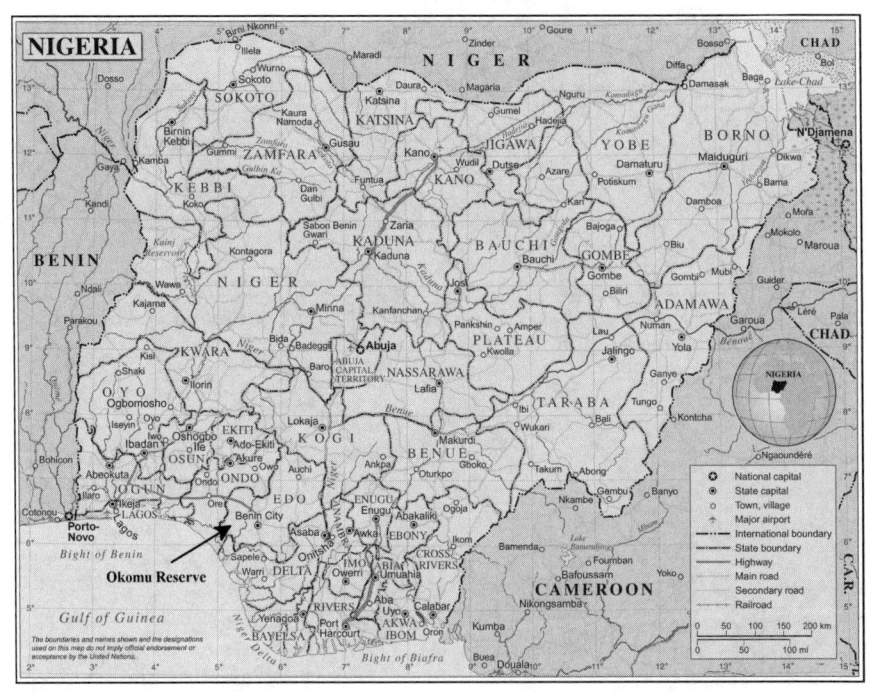

Map 1 Map of Nigeria, showing Edo State and Okomu Reserve (adapted from http://www.vidiani.com/?p=9544, downloaded 28 September 2012).

xiv | *Maps*

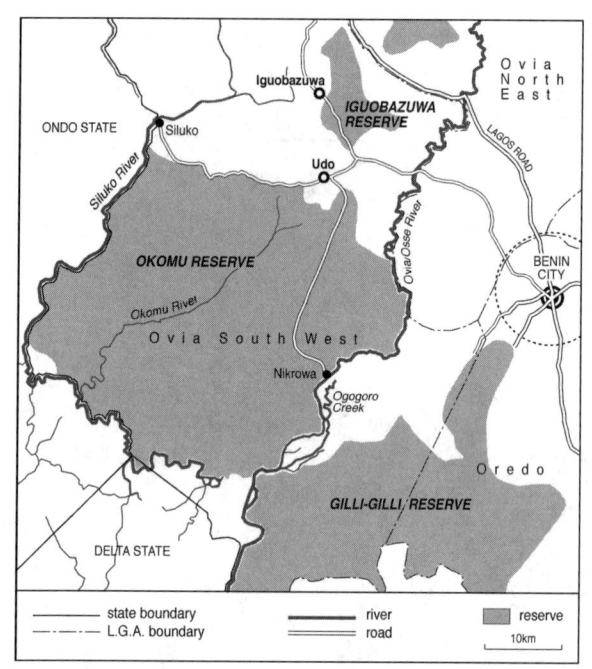

Map 2 Okomu Reserve in Edo State, Southern Nigeria (drawn according to my draft sketches by Hazel Lintott, 2005).

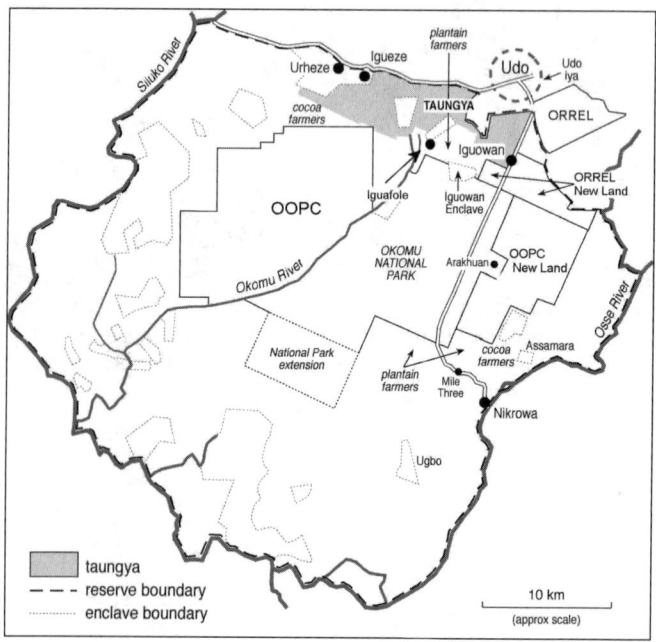

Map 3 Okomu Reserve (drawn according to my draft sketches by Hazel Lintott, 2005, with minor later revisions).

Introduction

In the colonial period the Benin Division of southern Nigeria, the heartland of the former Benin Kingdom, was the showcase of scientific forest management in West Africa. Its rich forests, which for years furnished the bulk of Nigerian timber exports, were protected by large-scale reserves, whilst logging activities and forest regeneration were regulated through comprehensive working plans. In recent decades, however, forestry in today's Edo State has been in a sorry state. Working plans and regeneration programmes have been abandoned, there is widespread illegal logging, and large parts of its reserves have been converted to farms and plantations. Forests are much reduced in size and timber resources more or less 'finished', as local logging contractors put it. Whilst Edo State's fall from its former pivotal place has been the most spectacular, forest management has similarly declined in many other parts of Nigeria, and deforestation is widespread.

According to international foresters and conservationists, there are clear links between recent management failure and deforestation in Nigeria. The *Conservation Atlas of Tropical Forests*, for example, states that Nigeria's 'natural forests were carefully managed in the early part of the century, [but] they have since been severely over-exploited' (Lowe et al. 1992: 230; see also Oates 1999). Nigerians themselves routinely blame corruption and greed amongst foresters and politicians for forest loss and the depletion of timber. Such accounts fit smoothly into the broader perception of Nigerian governments as corrupt and inefficient; into the constant refrain that 'things fall apart'[1] in Nigeria. But they are also symptomatic of a wider trend in the understanding of deforestation in Africa. For much of the twentieth century the main perpetrators in deforestation were seen to be small-scale farmers recklessly cutting down trees, either unaware of the environmental consequences of their actions or forced by economic necessity (e.g., Cleaver 1992). In recent decades, however, a critical body of research has shown that rural populations often actively improve rather than destroy their environments, and that the persistence of their negative perception presents a powerful environmental 'crisis narrative' that has been instrumental in justifying colonial and post-colonial land appropriation in the name of conservation (Fairhead and Leach 1996; Leach and Mearns 1996; Roe 1995; Spichiger and Blanc-Pamard 1973; Wood 1993). Instead, in Africa and elsewhere focus has shifted to the underlying, political causes of tropical deforestation (Contreras-Hermosila 2000; Geist and Lambin 2002).

This shift reflects in part the emergence of political ecology and its concern with the intersection of political economy and environmental change. Initially such political ecology work, such as Blaikie and Brookfield's seminal *Land Degradation and Society* (1987), was still primarily concerned with poor small-scale farmers, albeit highlighting how their practices were determined by larger political and economic processes beyond their control. But recently it has increasingly paid attention to deforestation caused by larger-scale agricultural projects, industrial logging and the various political processes behind this, such as Hecht and Cockburn's (1990) work in Amazonia, Klopp's (2000) study of elite land grabbing in Kenya, or Dauvergne's (1993) analysis of deforestation in Indonesia (see also Dove 1993; Rudel 2007).

At the same time, the shift reflects the rise of the 'good governance' agenda in international development since the 1990s. In the context of growing difficulties with the implementation of Structural Adjustment Programmes (SAPs), major donors and international financial institutions, such as the World Bank and IMF, increasingly identified 'bad governance' – inefficiency, corruption, weak legal systems and policing, lack of transparency and accountability – as root causes of SAPs implementation problems, and indeed of underdevelopment and poverty, and began to make lending conditional on the promotion of 'good governance' – efficiency, rule of law, transparency, accountability and participation.[2] Consequently the forestry sector has become increasingly concerned with the combat of corruption and in particular illegal logging, which have been identified as key causes of deforestation in recent years (e.g., Barbier Damania and Léonard 2005; Glastra 1999; Kishor and Damania 2007; Laurance 2004; Palmer 2001; Siebert and Elwert 2002).

Corruption, mismanagement and the non-implementation of official policies indeed often play a critical role in tropical deforestation, and need to be analysed and understood properly. Doing so not only helps to shift the blame for deforestation away from small-scale farmers, it also addresses a weakness in the existing critique of environmental crisis narratives. This literature readily acknowledges that policy processes are far from smooth and have many unintended outcomes (e.g., Fairhead and Leach 2003; Keeley and Scoones 2003; Leach and Mearns 1996). But because it focuses on exposing the continuities in conservation discourses and policies that deny farmers access to resource use, it pays less attention to actual policy practices and to the reasons behind the discrepancies between stated and actual practices. It is important to engage seriously with these discrepancies, in order to understand the actual outcomes of forest policies for both local livelihoods and landscapes.

However, in identifying corruption and governance failures as key causes of deforestation, and politicians and civil servants as its main agents, it is all too easy to be influenced by another set of crisis narratives, namely political ones. Political crisis narratives are in many ways just as pervasive, and instrumentalised just as effectively, as environmental ones. Accounts of corruption and mismanagement have, for example, played a key role in justifying the stringent SAPs of the 1980s and 1990s and good governance conditionalities since then (Harrison 2006; Szeftel 1998). They also play an important role in the internal politics of many developing countries. This is certainly the case in Nigeria, where political crisis narratives abound. Almost daily its newspapers feature articles entitled 'The problem with Nigeria', 'What's wrong with Nigeria?', or 'Things fall apart', and in everyday conversations many Nigerians discuss their political leaders and administrators in nothing but the most disparaging terms, with a constant circulation of tales of corruption, abuses of power and crime. Scholarship on Nigerian politics, too, tends to concentrate largely on how they are shaped by patron–client relations, corruption, graft, violence and predation (Eberlein 2006; Joseph 1983; Lewis 1996; Reno 2002). Overall perceptions are reflected in titles such as *Crippled Giant* (Osaghae 1998), *This House Has Fallen* (Maier 2000), or *The Criminalisation of the African State* (Bayart, Ellis and Hibon 1999), in which Nigeria features prominently.

Whilst there is much basis for such analyses, it is also the case that accusations of corruption and crime have long been used as political tools in Nigeria (Tignor 1993); they are themselves deeply engrained in Nigerian political culture and therefore cannot necessarily be taken at face value. Moreover, the entrenched, automatic evocation of corruption and government failure as the main causes of today's problems prevents more careful, nuanced analyses. This is the case with Edo State forestry, where deforestation is so routinely associated with recent mismanagement and corruption that there seems little need to investigate the current state of affairs further. Yet important as recent management failures are in explaining the current state of Edo forestry and forests, this analysis does not tell the whole story: there is more to recent developments than simply the collapse of a once properly functioning system of management.

This book exposes the insufficiencies of standard accounts of the collapse of forest management in Edo State in two ways. Firstly, it critically examines forest management itself. Underlying the identification of governance failures as the main causes of forest destruction, in Nigeria and elsewhere, is the assumption that 'proper' management does indeed protect forests; that scientific forestry's combination of forest reservation, logging regulations and regeneration programmes equates sustainable forestry. However, this cannot be automatically assumed. Throughout its history in Europe and the tropical world scientific forest management, even if properly implemented, encountered not only numerous political but also ecological problems, and did not always achieve its aim of sustainable timber production. These fundamental problems of forest management need to be taken into consideration when examining the managerial shortcomings of recent decades, as it matters how successful forest management actually was before its decline, and how its past history relates to present day management practices and environmental change. In Edo State, as this book will show, many of today's problems stem from past forest management itself, rather than from deviation from its principles.

Secondly, the book closely examines the nature and outcomes of recent developments. Given that patronage and corruption have indeed significantly shaped forestry practices in Edo State, it is important to understand their origins, manifestations and outcomes properly. Actual practices of today – such as the allocation of reserve land to political supporters, or the ways in which logging licences are granted – need to be related to the particular ways in which the forest economy became integrated into the political economy of the former Benin Kingdom over the course of the twentieth century. However, patronage politics do not explain all current practices. For one, in numerous instances forest officers and local communities have tried to resist land allocation or unregulated logging activities; not everybody participates in underhand dealings. Moreover, forestry has experienced many problems due to its internal shortcomings that were exacerbated by changing economic and demographic circumstances, but these problems were not addressed by official reforms or policy changes. In this context at least some of the ways in which forestry practices have diverged from official policy can be interpreted as unofficial forms of policy revision, developed informally on the ground in response to changing conditions. And finally, as this book will show, the environmental outcomes of these informal policy changes have not all been uniformly detrimental – some have brought environmental benefits with them.

A revision of existing accounts of the decline of Edo State forestry is important for our perception of contemporary Nigeria. Whilst the literature on Nigerian politics as a whole is very developed, there are few indepth studies of particular government sectors (Aluko 2006). Yet in order to discern the actual manifestations of patronage politics and everyday corruption, detailed, sector-specific ethnographic analyses can provide valuable insights (e.g., Anders 2004; Blundo and Olivier de Sardan 2006). Forestry is a useful sector to study in this respect as its role in resource control makes it particularly prone to corruption and mismanagement. A close investigation of forestry therefore allows us to properly understand such practices – their manifestations, causes and limits – and in this way to move beyond the routine crisis narratives that dominate perceptions of contemporary Nigeria.

At the same time, a reappraisal of Edo forestry provides valuable insights for forest conservation. Tropical forest governance has been the subject of wide-ranging reforms in recent decades, as the international good governance agenda and a shift towards decentralisation and community-focused approaches in development policy have coincided with renewed concerns over tropical deforestation and its effects on climate change. In the hope of making forest protection more equitable and democratic as well as more effective and sustainable, since the 1990s many countries in Africa and Asia have introduced reforms to decentralise state forestry services and to involve communities directly in forest management. In addition, initiatives such the Africa Forest Law Enforcement and Governance (AFLEG) programme under the umbrella of the New Partnership for Africa's Development (NEPAD) and, more recently, the United Nations Collaborative Programme on Reducing Emissions from Deforestation and Forest Degradation in Developing Countries (REDD) have sought to improve forest governance through international agreements.

Yet in many countries these reforms and initiatives have had only limited success: non-implementation and elite capture of forest profits remain as problematic as they were under central forest administration and reforms have not improved forest protection (German, Karsenty and Tiani 2010). There are many reasons for this limited success, but one I would like to suggest here is that forest reforms have often been undertaken without sufficient understanding of the reasons for the failures of centralised management and current environmental problems. With reforms focusing on implementation, the actual substance of forest management – forest reservation, logging and forest use restrictions, etc. – is rarely questioned and remains largely unchanged. As others have pointed out, communities are charged with more implementation responsibilities but not necessarily involved in decision making or the actual design of conservation policies (Adams and Hulme 2001b). In Nigeria itself, forestry has long been largely run by state governments rather than by the federal government, but further decentralisation towards the community level has not taken place; there have only been a few community conservation projects set up by NGOs in conjunction with wildlife conservation areas and National Parks. In this respect Nigerian forestry lags behind other African countries such as Tanzania, where the shift to community-based forestry is much further advanced. Nevertheless, a close examination of forestry in southern Nigeria still offers valuable insights to policy makers searching for more effective solutions today, as it allows us to properly understand the historical roots of recent problems of centralised forest management. Moreover, it shows that community conservation practices do not depend on the existence of official programmes, but can flourish quite independently, often in the very spaces created by the non-implementation of official policy.

A Political Ecology of Forest Governance: Place, Nature, History

The alternative reading of contemporary forest governance in Edo State presented in this book results from an analytical approach rooted in political ecology. Since Eric Wolf's (1972) original coining of the term 'political ecology' and early seminal texts such as Blaikie and Brookfield's *Land Degradation and Society* (1987), this interdisciplinary sub-field has developed in many directions, ranging from studies of rural resource conflict to deconstructions of global environmental discourses or concepts of nature (for overviews, see Biersack 2006; Peet and Watts 2004). The label 'political ecology' today embraces such diverse subject matters, theoretical leanings and research methods that it does not, as such, provide a distinct analytical framework or body of theories. Instead, discussions amongst political ecologists have centred on more loosely defined questions of theory and method, in particular around place, nature and history. In addition to its central concern with the intersection of political economy and the environment it is, arguably, an acute sensitivity to these questions that perhaps best defines a political ecology approach. In the following, the particular methodological choices that this study is informed by – an ethnographic focus on one locality; a serious engagement with ecology; and a deeply historical approach – will be discussed in relation to these wider debates within political ecology.

Place

One reason for the emergence of political ecology was a growing recognition of the limitations of cultural or human ecology, which tended to concentrate on human–environment relations within one locality (Rappaport 1967; Steward 1955; Steward, Steward and Murphy 1977). Political ecology's key contribution has been to relate local environmental change to wider national and global politico-economic processes. Nevertheless, many political ecologists continue to explore these wider processes through locally focused studies, and a sense of place remains crucial. Place here is 'not the local, not globality's Other, but, rather, the grounded site of local-global articulation and interaction' (Biersack 2006: 16). A particular strength of local studies is to understand the role of grass root agency in the local articulation of global processes and 'the perceptions, motivations and values that inform this agency' (ibid.: 26). The combination of different scales this involves can be challenging, and perhaps political ecology's emphasis on micro studies to some extent prevents larger, theoretical debates (Brown and Purcell 2005; Walker 2006). Nevertheless, locally grounded, in-depth analyses of policy practices and outcomes generate insights that are missed in studies researching policy processes on a larger scale.

This book's overall unit of analysis is the former Benin Division, the bulk of today's Edo State and the heartland of the erstwhile Benin Kingdom. It is the homeland of the Edo-speaking people. Indeed, the Benin Kingdom was locally known as the Edo Kingdom and its capital as Ubinu; the name Benin was given to kingdom, city, language and people (or Bini in the case of the latter two) by Portuguese visitors in the fifteenth and sixteenth centuries. Today 'Benin Kingdom' and 'Benin City' but 'Edo language' and 'Edo people' are the official terms used and have been adopted here. The area underwent numerous administrative changes after the collapse of the Benin Kingdom and its incorporation into the Niger Coast Protectorate in 1897 (see Appendix), but I am mainly referring to it here as the Benin Division because this was its name for most of the first half of the twentieth century (1914–1963) and as Edo State, its name since 1991. In this area the transformation of forestry – from intense forest management to its recent collapse – has been the most dramatic of Nigeria. But in order to really understand these developments they needed to be both related to larger developments in Nigeria and international forestry and explored in a more local context within Edo State. The locality I chose was the town of Udo and Okomu Forest Reserve in Ovia South-West Local Government Area (LGA), where I conducted eight months of ethnographic fieldwork between 2001 and 2003 and returned for one month in 2006 (Maps 1 and 2). It is located in the rainforest belt of southern Nigeria and has a flat terrain of typical Benin sands, with rainfall of about 1500–2000 millimetres a year. Okomu is a mixed, dry rainforest, dotted with small swamps (Jones 1955, 1956). Udo, on the reserve's northern border, is a small town today but was once a major rival of Benin City. This historic rivalry not only continues to play an important role in Udo's development and politics today, but it also significantly shaped the vegetation of Okomu Reserve. At 460 square miles, Okomu Forest Reserve is one of the largest reserves in southern Nigeria and was an important area for forest management and timber logging in the colonial period. Today it contains two large, expatriate-managed rubber and oil palm plantations as well as small-scale food crop, cocoa and plantain farms. It is also home to Okomu National Park, a wildlife conservation area that was first set up in 1985 (see Map 3). The Okomu area thus presents something of a microcosm of forest developments in Edo State as a whole, and was therefore an ideal location to explore these at a local level.

However, my research did not exclusively focus on Udo and Okomu. I also conducted interviews at the Forestry Department and at sawmills in Benin City, followed loggers and forest officers to other areas, and visited villages and forests throughout Edo State. My archival research, meanwhile, covered the Benin Division and forest management in southern Nigeria as a whole. The combination of these approaches enabled me to uncover not only local agency at the grass root level but also that of many other actors involved in Edo forestry at different stages of the policy process: colonial and contemporary foresters, regional and national governments, international conservationists, loggers and planters. In this way this study provides a holistic yet grounded sense of the overall development of forest policies in Edo State.

Nature

Diverse as the field of political ecology has become, there is a preoccupation with discourses, perceptions and constructions of nature, especially amongst anthropologists (Biersack and Greenberg 2006; Peet and Watts 2004). In part, this reflects a long-standing concern of anthropology, namely to discern and make intelligible different world views. But it also reflects the influence of constructivism, which problematises the concept of 'nature' and the idea of a 'real' nature independent of our perceptions of it (Demeritt 2002). Of course, an interest in the construction of nature does not necessarily preclude the idea of a 'real' nature out there; 'nature is simultaneously real, collective, and discursive – fact, power, discourse' (Escobar 1999: 2). But whilst many political ecologists do not actually question the existence of nature as such, they nevertheless tend to focus more on political economy than on nature itself; ecology is often secondary to politics (Vayda and Walters 1999).

This study's central concerns are policy processes and their role in the wider transformation of the landscape that occurred in the twentieth century; the emergence in southern Nigeria of what Escobar (1999) calls the 'capitalist nature regime'. Perceptions of nature also matter in this; as Escobar writes, the capitalist nature regime involved 'new ways of seeing, rationality, governmentality, and the commodification of nature linked to capitalist modernity'(Escobar 1999: 6). Here, both colonial foresters' attitudes to forest and those of local Edo people are discussed, underwriting as they did very different land-management practices. But they are not central to the analysis, as for example in Giles-Vernick's (2002) exploration of the concept of *doli* amongst the Mpiemu of the Central African Republic. Instead, the focus here really is on policy processes and their role in landscape change. For this reason, I sought to pay equal attention to political economy and ecology. Whilst I did not engage in systematic ecological research myself, I was able to closely observe vegetation in the course of many hours of walking through Okomu Reserve and surrounding areas in the company of local inhabitants, always intensely discussing the plants we saw in fields, forests and fallow lands.[3] I also researched older historical records and drew on existing archaeological and ecological research into Okomu's vegetation history, as well as the wider scientific literature on tropical and in particular forest ecology. This provided a sense of both longer and shorter term vegetation dynamics in the area, and of the role of policy in them.

In this approach to landscape change I have been informed by two new, related trends in ecology, namely disequilibrium ecology and historical ecology. Ecology as a discipline was initially built on concepts such as equilibrium, succession, climax and stability, which dominated the discipline in the early twentieth century (Anker 2001; Worster 1993). In the 1970s, however, something of a paradigm shift occurred when disequilibrium, instability and change increasingly began to be seen as the defining characteristics of ecology (Botkin 1990; DeAngelis and Waterhouse 1987; Hastings et al. 1993; Pickett and White 1985; Sprugel 1991). With recent research showing significant climatic changes, too, African forests and other ecosystems are no longer seen as stable but as naturally unstable, in constant 'flux' (Gillson, Sheridan and Brockington 2003).

At the same time, if ecology has traditionally focused exclusively on nature outside of human interference, there is now a growing recognition of the significant role that humans have long played in shaping environments throughout the world. Virtually all of the world's forests have experienced anthropogenic influences; it is increasingly obvious that there are no virgin forests today (Denevan 1992; Willis, et al. 2004; Yoon 1993). This longer term human history of forests is the subject of historical ecology, a field mainly dominated by anthropologists and archaeologists (e.g., Balée 1998, 2006; Crumley 1994) but also gaining ground within ecology itself (e.g., Bürgi, Russell and Motzkin 2000; Sheil, Jennings and Savill 2000). On the basis of a range of archaeological and documentary evidence, this study demonstrates that the Benin forests too were fundamentally shaped by humans. Overall, therefore, ecology is crucial to this analysis of forest policy and plays a central role in its critical re-evaluation of the dominant account of forest policy history in southern Nigeria.

History

Political ecology initially included little empirical historical research (Robbins 2004: 61) and history may still be 'one of the least visible' research areas within political ecology (Davis 2009: 285). Nevertheless, historical approaches are increasingly recognised as an important strategy in political ecology and a growing number of historical studies have made important contributions to the field (e.g., Bryant 1997; Davis 2007; Sivaramakrishnan 1999). Indeed, when trying to understand the intersection between environmental and political change, historical perspectives are crucial. For a start, recent processes of environmental change need to be related to longer term ones. In Edo State the longer term historical ecology of Benin forests has played an enduring role in twentieth-century forest dynamics. Similarly, the political and economic legacies of the Benin Kingdom influenced the ways in which forest policies were implemented and integrated into the local political economy. At the same time, the environmental and politico-economic changes brought about by the introduction of scientific forestry can only be appreciated through knowledge of existing conditions. For all these reasons it is necessary to place twentieth-century processes in their longer term ecological and political context, and to explore Benin and Udo's pre-colonial history in some depth.

But twentieth-century policy processes themselves also need to be approached historically. Policy processes are sometimes analysed rather formulaically, as a linear progression from formulation through implementation to outcome. In practice, however, they tend to be somewhat messier. In order to properly 'study through' policy (Shore and Wright 1997) it is crucial to appreciate temporal dimensions of policy processes and to study them over a longer period of time (Pierson 2004). Here, each policy – forest reservation, logging regulation, forest regeneration – is traced from its beginnings in the early twentieth century to the present day. This approach means that, as discussed before, it is policy itself that remains central, involving an ever changing set of actors along the way. This helps to discern the relations between past and present forestry practices and environmental change and to highlight the historical momentum behind recent developments.

In explicitly establishing twentieth-century continuities, this book not only follows a trend in the political ecology of forestry (e.g., Bryant 1997; Conte 2004; Sunseri 2009), but it is also part of a well-established tradition of Africanist research that explores the legacies of colonialism in the post-colonial African state (e.g., Chabal and Daloz 1999; Young 1988). This does not mean that Nigeria's independence in 1960 marks a key turning point in the narrative here. Indeed, twentieth-century African history can be conceptualised quite differently; Cooper (2002), for example, begins his study of modern African history in 1940, whilst Bernstein and Woodhouse (2001) identify different phases of development across the colonial and post-colonial period. This approach is taken here too, in order to put the particular turning point under discussion here – the decline of Nigerian forestry in the early 1980s – in its proper context.

Overall, then, the study's argument rests on tracing and emphasising historical continuities in both management practices and environmental change, in order to probe the perceived disjuncture between proper forest management in earlier parts of the twentieth century and its decline in recent decades. Nevertheless, this approach combines two distinct strategies: a critical assessment of past scientific forestry and an investigation into the role of patronage politics and corruption in recent developments in Edo forestry. The particular debates and literatures each strategy engages with need to be discussed in more detail, before briefly outlining the overall organisation of the book.

Scientific Forestry in Southern Nigeria

> Luckily for the future inhabitants of Southern Nigeria it has been decided not only that their forests shall be organised and looked after, but also that they shall produce timber and other minor forest produce ... My tours through the forests of Southern Nigeria have convinced me of their value to the native communities and ourselves, and of the great future before them if they are systematically organised on Indian lines.[4]

Scientific forestry was developed in eighteenth-century Germany. It was one of the cameral sciences (*Kameralwissenschaften*) that emerged in the course of a gradual reorientation amongst European sovereigns with regard to the overall aims and purposes of government, from sovereignty over territory to the regulation and growth of the economy. In his essay on governmentality, Foucault quotes the sixteenth-century writer Guillaume de La Perriere who aptly expresses this reorientation: 'government is the right disposition of things, arranged so as to lead to a convenient end' (Foucault 2000: 208). Foucault describes this 'convenient end' as 'a plurality of specific aims: for instance, government will have to ensure that the greatest possible quantity of wealth is produced, that the people are provided with sufficient means of subsistence, that the population is enabled to multiply, and so on' (ibid.: 211). In order to achieve these goals, prospective civil servants in eighteenth-century Germany were trained in different aspects of administration, public finance and economic development. Amongst these cameral sciences was scientific forestry, developed in order to manage and increase Germany's forest resources in the wake of growing concerns about diminishing timber supplies. Scientific forest management was subsequently taken up and developed further in Napoleonic France, where a forestry school was started at Nancy in 1824. When worries about deforestation amongst British colonial government officials in India induced them to take steps towards forest protection and management in the mid nineteenth century, they employed French and especially German foresters, notably Dietrich Brandis and Wilhelm Schlich, to set up a forest department and a system of scientific forest management (Rajan 1998, 2006). Forest departments subsequently established in other parts of the British Empire were mostly staffed by foresters trained in India (Barton 2001; Fairhead and Leach 1998). Ralph Moor, the Consul-General of the then Niger Coast Protectorate, initiated the establishment of a forest department in 1899, in order to 'increase the yield of existing known products by safeguarding them from damaging methods which might result in their extinction'.[5] In 1903 he appointed H.N. Thompson as the first Conservator of Forests of the Protectorate of Southern Nigeria, formed in 1900 through the joining of the Niger Coast Protectorate and the territory of the Royal Niger Company. Thompson had twelve years of experience in setting up the Forest Department in Burma, and had 'thus been able to pass through and become familiar with all the necessary stages in forest organisation from its introduction to its completion.'[6]

Scientific forestry epitomised the ethos and vision of modern state government. It had itself a 'plurality of different aims': mainly to ensure long-term timber supplies and to create government revenue, but also, especially from the nineteenth century onwards, to contribute to watershed protection and climatic stability (Grove 1995). These different developmental and environmental goals were to be achieved through scientifically designed working plans. Working plans regulated logging activities and tree regeneration programmes in government-controlled forests, resulting in the systematic transformation of forests into more regimented collections of trees nearing foresters' ideal of the *Normalbaum* (the 'normal' tree) (Lowood 1990). As such, early scientific forestry served as 'something of a model' to James Scott (1998: 11) for the processes he analysed in *Seeing Like a State,* his study of large (and failed) state-led projects of rural and urban transformation.

In the colonial context in particular, scientific forestry has been analysed in these terms. Bryant describes how in Burma, as in Germany before, scientific foresters attempted to rationalise forests in the Weberian sense of 'formal rationalisation', writing that 'this process of enhanced control over nature and people through non-revolutionary change is nowhere more evident than in the doctrine of scientific forestry'(Bryant 1998: 829). Sivaramakrishnan (1999) examines forestry in northern India as a form of 'state making', presenting as it did one of the most tangible manifestations of colonial governmentality, whilst Vandergeest and Peluso (2006a, 2006b) examine forestry in Southeast Asia as a form of state power and territorialisation. A different yet no less important aspect of governmentality, namely the gradual internalisation of official conservation discourses and practices by local populations, is explored by Agrawal in *Environmentality* (2005). Worboys, finally, discusses scientific forestry as one of the 'imperial applied sciences' that were central to Chamberlain's policy of 'constructive imperialism' (Worboys 1979, 1996).

Too great an emphasis on scientific forestry's representativeness of modern governmentality, however, detracts from the many difficulties encountered in the course of its implementation, as well as from its unintended outcomes. For a start, in the colonial context (but also in Europe) forestry rules evoked considerable resistance from local populations. Just as colonial forestry was tangible 'state making', profoundly transforming subjects' lives, so protests against it were one of the main forms of resistance against colonial governments, with colonial subjects in South Asia and Africa starting fires, disobeying forest regulations, demonstrating and petitioning against forest reservation and restrictions on forest use (Agrawal 2005; Anderson and Grove 1987b; Grove 1990; Guha 1989; Peluso 1992; Sunseri 2009). But there were also more internal problems. Financial and other constraints meant that many forestry policies were delayed by several decades, were run with minimal resources, or had to be abruptly abandoned. In this sense forestry was symptomatic of the often makeshift nature of the colonial state in Africa (Berry 1993; Phillips 1989). Chabal and Daloz (1999) point out that colonial rule in Africa was determined by pragmatism and ad hoc solutions rather than the inculcation of new political habits. They describe how French district officers 'tended to discharge their duties in a personalised, arbitrary and "unofficial" manner, which ill attributed to the development of modern bureaucratic order' (see also Blundo and Olivier de Sardan 2006: 43; Chabal and Daloz 1999: 12). Forestry was in this respect perhaps particularly prone to abuse. From its beginnings in Germany and France, the regulation of logging activities through licenses and government controls left wide scope for abuse by forest guards and officers, whilst illegal logging and underhand dealings were always difficult to control.[7] In general, the manner in which new colonial policies, such as forestry, became part of the local political economy rarely resulted in straightforward rural development, but often provided more economic opportunities for political elites. Osoba writes that in Nigeria, 'the colonial authorities and their collaborators presided over a fraudulent and corrupt accumulation system' (Osoba 1996: 373).

At a higher decision-making level, too, colonial forest departments experienced problems. There were frequent conflicts of interest with other government departments, for example with Native Affairs in South Africa (Tropp 2006) or with Agriculture in the Central Provinces of India (Rangarajan 1998). There were also conflicts with commercial logging companies (Anderson 1987). Forest departments often had a somewhat ambivalent attitude towards logging companies, wishing to both foster and control logging activities; in this respect, forestry revealed some of the 'inherent contradictions' of colonial rule (Berman and Lonsdale 1992). In general, the balancing of forestry's developmental and environmental goals – developing a large timber industry but also ensuring long-term timber regeneration and, in some places, watershed or soil protection – was extremely difficult in practice.

Most fundamental of all, the fact that colonial forestry consisted of the application of a standard package of measures – forest reservation, logging regulation and either natural regeneration or artificial regeneration through planting – made adaptations to local conditions more difficult. This is not to say that colonial foresters necessarily misunderstood forest ecology – on the contrary, in the course of their work they gained many insights and at times in-depth knowledge. But rationalisation through centrally devised policies was so central to scientific forestry that there was little scope for such local insights to lead to fundamental revisions of approaches (von Hellermann 2011). This meant that some forest policies, even when properly implemented, did not have the results hoped for because they were not appropriate for local conditions.

Here, then, I will explore the ways in which scientific forestry in the Benin Division diverged from successful, rational resource management in both practices and outcomes. Its outcomes will be discussed primarily in relation to scientific forestry's own goals, namely forest protection, the development of a sustainable and profitable timber industry (to create revenue and foster local development) and, central to all this, the regeneration of economic timber species. But in order to discern scientific forestry's overall legacy, outcomes will also be conceived of more broadly, considering the larger environmental and social changes brought about by different forest policies. With regard to environmental changes, timber regeneration is not the only concern of foresters and conservationists today; biodiversity and – because of forests' potential role in ameliorating climate change – biomass and carbon sequestering have also become key criteria. With regard to social changes, I consider policies' general effects on local livelihoods but focus in particular on the ways in which they became part of and reshaped existing politico-economic structures. Both environmental and social changes resulting from colonial forest policies' have played a significant part in more recent developments.

Forest Governance and Politics in Contemporary Nigeria

The collapse of forestry in Edo State and elsewhere in Nigeria has been part of a wider decline of the country's government and infrastructure. Whilst the discovery and exploitation of oil from the late 1960s initially brought revenue and wealth, this oil-fuelled growth came to an abrupt end in the late 1970s, when falling production and oil prices triggered an economic collapse. In the 1980s reduced oil revenue combined with Structural Adjustment Programmes, under Ibrahim Babangida's military rule, to bring about a drastic reduction in government spending on infrastructure and a wide range of government sectors (Forrest 1993; Osaghae 1998). In forestry this involved large-scale staff retrenchment and the virtual cessation of any investment. However, global economic crisis and structural adjustment alone did not account for all economic and administrative problems: as is well known and much discussed, Nigerian governance under both military and civilian regimes has been characterised by neopatrimonialism, corruption, rent-seeking and increasingly predatory rule (Eberlein 2006; Joseph 1983; Lewis 1996; Reno 2002).

If Nigeria is perhaps particularly well known for such political practices, they are also widespread in many other African countries. In the literature analysing post-colonial African politics, two broad explanatory approaches have emerged: one focuses on African political traditions and attitudes towards power, the other on the legacies of the colonial state. Key amongst those exploring the more endogenous aspects of the contemporary African state is Bayart (1993), who famously described African politics as 'politics of the belly', where power and access to wealth are perceived as virtually interchangeable. This attitude is expressed in the frequent use of metaphors of food in West African political culture, such as, for example, in the Cameroonian saying 'a goat grazes wherever it is tied up'. With regard to corruption specifically, the anthropologist Olivier de Sardan (1999) identifies five logics 'profoundly ingrained in current social life [in Africa]', influencing both big-time and petty corruption. These are the logics of negotiation, of gift giving, of the solidarity network, of predatory authority, and of redistributive accumulation. Olivier de Sardan is keen to emphasise that this does not mean that corruption is intrinsic to African culture; rather, he argues, these existing practices favour the 'generalization and banalisation' of corruption (see also Anders 2004; Smith 2001, 2007). In general, such 'moral economy' explanations feature prominently in the literature on corrupt practices in Nigeria (Aluko 1966; Ekpo 1979; Enahoro 1966; Joseph 1983).

In contrast, others emphasise the legacies of the colonial state in shaping current political practices. As discussed in the previous section, insufficient resources, the personalised manner in which some colonial officers governed and the fact that the integration of local elites into the colonial administration provided many opportunities for personal accumulation all meant that patronage politics, rent-seeking and corruption already emerged in the colonial periods. Njoku (2005) and Munyae and Mulinge (1999) argue that indirect rule corrupted existing indigenous practices such as gift giving, which previously were expressions of solidarity and democracy rather than corruption. Others have focused on people's actual experience of the colonial state, arguing that it was perceived as far outside society, alien, imposed, associated with the 'white man', and not deserving of citizen obligations and duties (Osaghae 1998: 21). Ekeh (1975) speaks of 'two publics', and Mamdani (1996) of the bifurcated or Janus-faced state, which operated differently in urban and rural areas. Taking a slightly different tack again, Sara Berry (1993) points to the larger organisation of many colonial economies, in particular the heavy reliance on marketing boards, which meant access to wealth was largely restricted to those with state positions. The 'overdeveloped state' argument then also served to justify Structural Adjustment Programmes in the 1980s (Szeftel 1998).

These two approaches are often portrayed as opposite: 'either the modern state is corrupted by traditional culture, or traditional culture is corrupted by the advent of the modern state' (Blundo and Olivier de Sardan 2006: 29). It is, however, unhelpful to take either approach to extremes – either seeing a 'misplaced allegiance to traditional social ties and gift economies' (Robbins 2000: 425) as the key reason for corruption, or painting a picture of an unspoiled, political Eden in pre-colonial Africa. It is more useful to consider how different politico-economic organisational structures and practices have developed alongside each other from the colonial period to the present day, and how these have resulted in discrepancies between stated objectives and actual practices. Such an approach was taken, for example, by Sharpe (2005) in his study of the role of forestry in the local political economy in Cameroon and by Grainger and Konteh (2007) in their long-term analysis of forest policy in Sierra Leone. It is also the approach taken here, in that patronage politics and corruption in Edo forestry today can be traced back to the particular ways in which the forest economy developed in the former Benin Kingdom, where economic organisation was largely structured by the kingdom's patrimonial hierarchies.

However, it is important to stress that patrimonialism, rent-seeking, corruption and predation are not the only logics of contemporary African politics and governance; the focus on these in the literature is in many ways symptomatic of the 'crisis narratives' shaping our perceptions of Africa. A growing body of literature now critiques the predominance of negative assessments of African politics (Mustapha 2002) and highlights the importance of popular ideas of accountability and legitimacy that challenge abuses of power (Hagberg 2002; Kelsall 2003). In Nigeria in particular, a number of recent studies have explored the ways in which ordinary people resist corruption, crime and predatory rule and seek to gain some control over the running of their affairs (Gore and Pratten 2003; Meagher 2007; Pratten 2007; von Hellermann 2010). Within African administrations, too, there are many different actors who are differently motivated and draw on a multiplicity of legal and moral frameworks (Anders 2004; Blundo and Olivier de Sardan 2006). In Edo forestry, as this study shows, both ordinary citizens and some administrators seek to uphold proper administrative procedures and to actively resist illegal and corrupt practices.

Corrupt practices themselves, moreover, are not all the same: the differences between large-scale and petty corruption, and the range of practices that can exist within one institutional or regional context, must be appreciated. In-depth ethnographic studies can make an important contribution to gaining a more nuanced understanding of this (ibid.). In particular, I suggest that some 'corrupt' practices in Edo State forestry may be understood more usefully as policy adaptation, rather than purely as abuse of office. This argument was made several decades ago by James

Scott, who suggested that, where interest structures and institutionalised forms through which demands can be made are weak or non-existent, a sizeable number of demands reach the political system after laws are passed, at the enforcement stage. Thus he writes that 'the peasants who avoid their land taxes by making a smaller and illegal contribution to the disposable income of the Assistant District Officer are as surely influencing the outcome of government policy as if they formed a peasant union and agitated for the reduction of land taxes' (Scott 1969: 326). Whilst Scott was mainly writing about Southeast Asia, similar arguments can be made in many African contexts, where official policy reforms are often slow and not necessarily informed by democratic processes. Anders (2004) describes how many official procedures in the Malawian civil service have remained unchanged in decades, and the same striking continuity of practices can be observed in Nigerian forestry. Logging allocations, for example, are broadly made in much the same way as when logging regulation started at the beginning of the twentieth century. In this context some current forest management practices may best be interpreted as strategic alterations that adapt the system to present-day demands. Rather than as corruption per se, they are instances of policy being made 'on the ground', in the interaction between local populations and 'street-level bureaucrats' (Lipsky 1979). Indeed, some of the new practices developed on the ground in Edo State were eventually made official policy, a process also described by Mosse (2004). I further suggest that it is in these unofficial, hybrid spaces that rural populations have regained some control over land management and that here one may find actual community conservation practices.

Finally, the environmental as well as social consequences of deviation from official conservation policies – through resistance, corruption, or lack of investment – need to be considered; indeed, they are now emerging as an important area of interest for political ecologists (Corbridge and Kumar 2002; Robbins 2000, 2004; Robbins et al. 2007; Robbins et al. 2009; Robbins et al. 2006). This study shows that in Edo forestry not all deviations from official policy are uniformly environmentally destructive; some even have positive environmental outcomes and contribute to local economic development. Just as scientific forestry was not as successful as it is remembered today, so recent changes have not been as detrimental as usually assumed. At the same time, recent non-implementation and divergence from formal policy have brought many opportunities but also risks and conflict for people living in the Okomu area – all these need to be considered in depth in order to gain a fuller appreciation of the overall effects of non-implementation.

Organisation of the Book

Having discussed past scientific forestry and contemporary forest governance separately here, it is the continuities between them that this study seeks to highlight. For this reason the book is organised thematically, with three core chapters each tracing one policy area from its beginnings to the present day. Each of these challenges standard accounts of the history of Nigerian forestry – that proper management and effective forest protection gave way to management failures and deforestation – in different ways. After an introduction to the research area and its longer term environmental and political history in the next chapter, Chapter 3 is concerned with the spatial transformation of Benin landscapes through reservation and then dereservation. It argues that reservation was not simply forest protection; rather, it disrupted the existing farm–forest symbiosis in which many timber species thrived and itself created the system of centralised land control that enabled rapid dereservation and forest conversion in recent decades. Chapter 4 investigates forest management, namely logging regulation and tree regeneration programmes. It shows that these never amounted to sustainable forest management and that, therefore, the collapse of working plans in recent decades must be understood as a final phase evolving out of previous developments, rather than as an abrupt departure. Chapter 5 brings together themes of the previous chapters – changing forest–farm relations and colonial timber-regeneration programmes – by looking at Taungya farming, an agro-forestry method introduced to Nigeria in the 1920s in order to simultaneously foster tree regeneration and address land shortages. In the case of Taungya, recent divergences from official policy have so many positive environmental and social outcomes that it can be interpreted as successful adaptation rather than management failure. Together, these three chapters present a powerful critique of standard accounts of the decline of forest management in Edo State, which is summarised in the conclusion to Chapter 5. The final, sixth chapter takes a different approach and may best be understood as a postscript to the book's core analysis. It is concerned with Okomu National Park, a conservation area first established in the heart of Okomu Reserve in 1985. It shows how a thorough understanding of the longer term history of forest management and local forest relations provides valuable insights when assessing recent conservation and community oriented approaches and concludes by briefly reflecting on the broader implications of this study for future forest policy.

Notes

1. The popularity of the phrase 'things fall apart' in Nigeria is, of course, due to Chinua Achebe's famous novel *Things Fall Apart* (Achebe 1958). While Achebe's book is concerned with the impact of colonialism, the phrase is now often used to describe the country's economic and political woes in recent decades.
2. 'What is Good Governance?' Retrieved 18 July 2012 from http://www.unescap.org/pdd/prs/ProjectActivities/Ongoing/gg/governance.asp. For a critical appraisal of the good governance agenda, see Weiss (2000).
3. To take Ingold and Vergunst one step further (so to speak), I would argue that we do not only need an ethnography of walking, but also more of an appreciation of walking as an ethnographic method (Ingold and Vergunst 2008).
4. The National Archives (NA), CO 879/69, Enclosure 3 in No. 138, Address by the Conservator of Forests, Southern Nigeria, before the Chamber of Commerce, Liverpool, 16 September 1904.
5. Nigerian National Archives Enugu (NAE), CSO 3/5/3, Sir Ralph Moor, H.R.M. Commissioner and Consul General to the Secretary of State for Foreign Affairs, 14 November 1898.
6. The National Archives (NA), CO 879/69, Enclosure 3 in No. 138, Address by the Conservator of Forests, Southern Nigeria, before the Chamber of Commerce, Liverpool, 16 September 1904.
7. This emerges, for example, from an eighteenth-century painting in the town hall of Raon l'Etape in France, which depicts four forest officers supervising the logging of one tree timber – four being necessary to check on each other and prevent corruption (Rochel 2005).

CHAPTER 1

Ecology and Politics in the Benin Kingdom

Introduction

The Benin Kingdom is commonly known as one of West Africa's intriguing forest kingdoms (e.g., Roese and Bondarenko 2003). Benin, Asante and the southern Yoruba city states were all situated in today's 'forest zone' along the West African coast and are historically interesting because political centralisation in West Africa is generally associated with the open savanna areas further north, where it would have been easier to maintain control over large areas and populations. It has therefore been a challenge to historians and archaeologists to understand and reconstruct the process of state building in the rather hostile environment of a rainforest (e.g., Connah 1987; see also McCann 1999; Wilks 1993). However, there is now increasing evidence that today's forest zones may not always have been covered in forest. Findings from palynological, hydrological and anthracological research indicate a more turbulent vegetation and climate history of the region than had previously been assumed. There is a large amount of evidence of severe dry periods affecting different parts of West Africa between 4,000 and 2,500 years ago, which still exert a major influence on vegetation patterns today (Maley 1996, 2002; Vincens et al. 1999). Their most lasting effect was the formation of the Dahomey Gap, a savanna corridor interrupting the forest zone that was often cited as an example of human-induced vegetation change (Salzmann and Hoelzmann 2005).

There is also evidence of less pronounced dry periods in the more recent past, in particular from Lake Bosumtwi in Ghana (Brooks et al. 2005; Shanahan et al. 2006; Talbot and Delibrias 1977). Historical records support the notion that there were drier and wetter phases in West Africa during the last few centuries (Brooks 1993; Nicholson 1978, 1980). On the basis of observations made by travellers to West Africa, Fairhead and

Leach (1998) show that along with climate, the extent of forests changed. They demonstrate that the forest zone west of the Dahomey Gap, far from being consistently covered in forest, had a varied vegetation history, and that large areas covered in forest in the early twentieth century were more open in the past. This was not just due to climatic influences but also to human history; seventeenth-century travellers' reports, for example, describe many parts of the West African coast as considerably more densely populated than they were two centuries later.

This chapter demonstrates that the Benin Kingdom too was more open and had a more dynamic vegetation history than the term 'forest kingdom' suggests. This matters for our understanding of the vegetation history of West Africa – putting twentieth-century deforestation statistics in some perspective – and of the development of the Benin Kingdom, perhaps less wondrous in a mixed landscape than in thick rainforest. But here, the main purpose of presenting Benin's long-term ecological history is to provide a sense of the particular composition of the Benin forests, shaped by centuries of farming, warfare and population changes, which came under scientific management in the twentieth century. At the same time, the chapter explores the political and economic history of the Benin Kingdom and of Udo Town. This history is important not only with regard to its role in shaping forests, but also with regard to the politico-economic context in which forest policies unfolded in the twentieth century. The chapter also serves as an introduction to Udo Town and its surroundings, the area historically known as Iyek'Ovia. Here, it focuses in particular on people's relation and uses of land, in order to understand not only the role of land-use practices in the shaping of landscapes, but also the changes and continuities in land use that occurred in the course of the implementation of twentieth-century forest policies.

The chapter begins, then, with a section on the early environmental and political history of the area, ranging from about 4000 BP, when the first Edo ancestors settled in the area, to the early sixteenth century, the time when Udo was finally defeated and fully incorporated into the Benin administration. This draws on existing archaeological, palynological and ecological research as well as local and regional historiography. This historiography, especially that of Udo, is largely based on oral tradition, difficult to verify, and frequently contested. Aware of its potential inaccuracies, of mythologised rather than factual representation of events, and of the politics behind different accounts, I nevertheless present it here as more or less the actual course of events. This is not only because of the absence of other sources, but also because it is highly meaningful to Edo people today.

The following section, broadly covering the sixteenth to early nineteenth century, draws on the descriptions of European visitors – again far from reliable sources, as will be discussed below – to convey a sense of Benin landscapes and land use at the time. It also explores in more detail the political and economic organisation of the Benin Kingdom, in particular its elaborate chieftaincy structure and the *obas'* extensive patrimonial powers. Patrimonial rule and political centralisation were, however, always counterbalanced by relative autonomy in village governance. Village life and land-use practices in Iyek'Ovia are focused on in the third section. The chapter ends with a brief discussion of the Kingdom's political decline in the nineteenth century and the environmental changes it brought with it, resulting in the 'rich' Benin forests encountered by colonial foresters after the Kingdom's defeat in 1897.

Early History

There is some evidence that southern Nigeria experienced the same dry period that affected the rest of West Africa between 4000 and 2500 BP. Thurstan Shaw's excavations at Iwo-Eleru near Akure (68 miles northwest of Benin City) revealed gloss-edged trapezoid stone tools appearing around 4000 BP, together with a marked increase in querns and rubbing stones used for grinding cereals, which could potentially be associated with cereal production and therefore a drier, more open environment (Shaw and Daniels 1984: 55–56),[1] whilst Russell Barber's analysis of land snail remains at nearby Igbo-Iwoto Esie suggests that this area was savanna-like and only became more forested around 1200 BP (Barber 1985). Moreover, palynological research in the Niger Delta has shown that a sudden spread of oil palm in the Niger Delta occurred between 4000 and 3000 BP (Sowunmi 1986). This was originally interpreted by Sowunmi as the result of human forest clearing, but is now regarded rather as natural colonisation by pioneer trees after a dry period, with their expansion most likely aided by humans (Maley 2001; Sowunmi 1999). Indeed, linguistic evidence suggests that migration of Edo ancestors from the Niger-Benue confluence into their present area took place between 4,000 and 3,000 years ago (Darling 1977, 1984).[2] According to the Edo myth of origin, Edo land was created by the youngest son of the high god Osanobua, who poured sand out of a shell, which has been interpreted as the mythical description of the onset of a dry period (Imafidon 1987). In fact, a dry period may have been the cause of migrations into the southern parts of Nigeria at this time (Andah 1987). For a people relying primarily on yam and palm oil for food,[3] conditions further north may have become too dry, whilst in the South a more open landscape would have been a more attractive one to move into than high forest or swamps. As light-demanders, neither yam nor oil palm grow naturally in high forest, and the spread of yam and palm oil cultivation in high forest over a thousand years before the development of iron technology (around 1500 BP) has long been a conundrum for archaeologists and historians (Andah 1987, 1993; Bondarenko and Roese 1999; Darling 1977; Shaw 1980). A more open environment makes the spread of agriculture in the Benin area and other parts of today's forest zone at this time less difficult to explain – indeed, similar arguments have been made recently about the Bantu expansion, vegetation change and population movements in southern Cameroon (Ngomanda et al. 2009).

The beginnings of the Benin Kingdom itself are somewhat shrouded in mystery, but have been dated to around AD 900 (Egharevba 1968). The first three hundred years or so of Edo history are known as the Ogiso period, with a series of mythical Ogiso rulers. At this time Benin's power did not extend far, and several Edo towns and villages, whilst ethnically, linguistically and culturally the same as Benin City, were still independent of Benin. Udo was one of these independent Edo settlements. It was founded at around the same time as Benin City, around forty miles northwest from Benin on the other side of the Ovia river, and was then a town of about equal size. According to oral history, several of the early Ogisos came from Udo, and it is possible that the two towns were originally competing to become the capital of the kingdom, a conflict eventually won by Benin City (Darling 1984; Ezele 2002b; Imafidon 1987).

The Ogiso period came to an end when the last Ogiso ruler did not have a successor, and the people of Benin asked the king of Ife, the most important Yoruba town, for help.[4] He sent his son Oranmiyan, who is said to have come to Benin City via Udo. Oranmiyan eventually returned to Ife, but his son by an Edo woman, Eweka, became the first *oba* (king) of Benin, the first in a long line stretching all the way to the current thirty-seventh *oba*, Oba Erediauwa.[5] The archaeologist Patrick Darling notes that many of the early *obas'* mothers came from villages surrounding Udo. This too may suggest that Udo played an important, if ambivalent, part in the emergence of a Yoruba style, divine centralised kingship in Benin (Darling 1984).

During the reign of Oba Eweka and his immediate successors, Benin slowly expanded in size and power. Eweka created the most important class of Benin chiefs, the 'kingmakers' or *uzama nihinron*, consisting of the *oliha, edohen, ezomo, ero, eholo n'ire* and the *oloton*. These had hereditary titles, were in charge of various Edo shrines and had the important role of crowning each new *oba* (Egharevba 1968).[6] Udo, at this time, was ruled by men remembered as 'powerful heroes', such as Ighoghogiamu, Obaghalakhara, Esu-Uhumwun and Owenbedeku (Ezele 2002b: 9). The last and most powerful of these was Akpanigiakon. He is reputed to have dug the *iya*, the moat and earth wall still surrounding Udo today, second only in size only to that of Benin City, which was built around the same time. Udo's population was then relatively high and there were many smaller farming settlements surrounding it. Concrete evidence for this has come from within Okomu Forest. In 1948 a Cambridge botanical expedition led by E.W. Jones came to Okomu to study a large undisturbed West African rainforest. However, a series of pits dug in the heart of the reserve for soil analysis almost all contained fragments of pottery and pieces of charcoal, many of which were recognizable as pieces of oil palm kernels (Jones 1956: 101). Jones speculated that 'their abundance certainly points to the former existence of a country very different from the forest now existing – to something like that which prevails in parts of the Owerri district east of the Niger, perhaps where there are no large villages, but groves of oil palm interspersed with groups of huts and small patches of cultivation cover large areas' (ibid.: 102). In 1998 John Oates and Lee White took two further charcoal samples in Okomu Forest Reserve and had them analysed. Radiocarbon dating revealed charcoal from the two sample pits to date from around AD 1100–1300, round about the time of Akpanigiakon and the building of the Udo *iya*. The charcoal collected from both pits again consisted almost entirely of fragments of oil palm kernels, which together with isotope analysis of the soil samples suggested to White and Oates that 'the landscape at the time that the charcoal and pottery layer was created would have been a mosaic of farmland, of palm trees, and secondary forest, dominated by C_3 plants' (White and Oates 1999: 357). They believe that an area of around 400 square miles may have been under intense agriculture.

A large and populous Udo and its powerful leader Akpanigiakon certainly presented a real challenge to Benin, then ruled by Oba Ogulua. There was consequently much intrigue in Benin against Akpanigiakon and frequent warfare between the two towns. In the end Akpanigiakon was killed at Urezen (a village further west of Udo to which he fled) by Ogiobo, a general in the Oba's army. In gratitude, Oba Ogulua gave Ogiobo Akpanigiakon's crown, making him the *onogie* (prince) of Udo. The *enogie* (princes), many of whom were sons of *obas*, played an important role in the administration of the Benin Kingdom, with a number of its towns and villages under the control of different *enogie* (Bradbury 1957; Eweka 1992). The creation of the *Onogie N'Udo* title meant that Udo, and Iyek'Ovia around it, now formed a province under the control of Benin. Several subsequent *obas* made their sons *Enogie N'Udo* (Egharevba 1968; Imafidon 1987).

The reign of Ewuare in the mid fifteenth century, the first and most legendary of the 'great' *obas* of Benin (Ewuare, Ozolua and Esigie), signifies the beginning of the period of Benin's greatest power. Ewuare expanded his kingdom's territory through military campaigns, developed the capital city and built up its chieftaincy structures. In particular, in order to curb the powers of the *uzama*, he created new, non-hereditary groups of chieftaincy titles: the *eghaevbo n'ore* or the town chiefs, headed by the *iyase*, and the *eghaevbe n'ogbe*, the palace chiefs, headed by the *uwangue*. The *iyase* was the most important town chief as well as warlord, whilst the *uwangue* was in charge of looking after the king's personal wardrobe and regalia. However, Ewuare's administrative reforms did not prevent upheaval and conflict, which continuously re-emerged within Benin City and throughout the Benin Kingdom (Egharevba 1968; Okpewko 1998).

Conflict between Udo and Benin broke out again in the early sixteenth century, at the time of Oba Esigie and his brother, Aruaran of Udo, who were both sons of Oba Ozolua. Aruaran is the most famous of all Udo rulers and still central to its identity today (Fig. 1). According to oral tradition, Aruaran was born just before his brother Esigie but did not cry immediately after his birth; because this was the way in which birth was announced, Esigie was declared the first born and official heir to Ozolua. Ozolua may, however, have secretly favoured Aruaran. He not only made him *Onogie N'Udo*; when he did so, he sent him to Udo with a large group of Benin chiefs accompanying him. These chiefs, it is said, resisted at first, but in the end had no choice as Ozolua forced them to *debavboghayu Udo* (join the other chiefs going to Udo) (Ezele 2002b: 5). According to Jonathan Ezele – the last *Oliha N'Udo* and the town's local historian – new people had to fill the now vacant chieftaincy positions in Benin City. At Udo, '[Aruaran] planted his father's and Ewuare's shrine and other departed *obas*' shrines so that the annual Ugie festival could take place as it was done in Benin by the chiefs ... The *oba* made a law that from time hence forth all the shrines of the departed *obas* should be erected at Udo' (Ezele 2002b: 5). Benin's guilds, town quarters (idumu) and street names too were created at Udo.[7] These developments amounted to an open challenge to Benin City, and eventually, after Ozolua's death, war broke out between Aruaran and Oba Esigie. There were several fierce battles, the worst of which is known as 'the battle of blood' (*Okuo-ukp-Oba*). Whilst the overall course of events is disputed, the war ended when Aruaran drowned in the lake of Udo, Odighi N'Udo.[8]

After a few months the *Iyase N'Udo*, Osemwughe, declared a new war against Benin to avenge his master's death, whereupon Esigie sent troops to Udo and several battles took place. This time Udo town itself was completely destroyed, and Osemwughe and most Udo inhabitants fled to other parts of southern Nigeria.[9] After Udo's destruction, Oba Esigie decreed that from then on no more *enogie* would be sent to rule Udo; instead, there would be only *Iyases N'Udo*. According to Ezele (2002b: 9) he also decreed that the *iyase* should render service to all the *oba* shrines and to hold all the annual celebrations honouring departed *obas*, such *erinmwindu*, *ikhure* and *igue-ugie*.[10] With the defeat of Aruaran, Udo had finally been subjugated and become a subordinate part of Benin. The importance of the defeat of Udo is reflected in the festivals celebrated in Benin City commemorating this event each year (Ben-Amos 1995; Darling 1984; Imafidon 1987).

Figure 1 Statue of Aruaran in the centre of Udo. Photograph taken by the author, 13 November 2002.

The Benin Kingdom from the Fifteenth to the Eighteenth Century

Following the large exodus of people from Udo, the farmland surrounding it was abandoned and eventually turned into one of the largest forest areas of the Benin Kingdom, today's Okomu Forest. Other parts of the kingdom that were deserted after warfare similarly became more forested, so that Benin's landscape as a whole reflected its turbulent history (Okpewko 1998). Generally, however, the Benin landscape remained quite open, with substantial areas under agricultural use and relatively high population levels. As in other parts of the West Africa forest zone, which were also quite densely populated at this time (Fairhead and Leach 1998), this becomes apparent from the records of European visitors who started coming to Benin from the late fifteenth century onwards. The first of these were the Portuguese, who sought to expand their naval empire, to trade and to spread Portuguese culture and religion in West Africa (Blake 1942; Crone 1937; Pereira 1937). The Portuguese played an important role in Benin; in fact, it may well have been Oba Esigie's ability to make use of Portuguese firearms, or even soldiers, that helped him to finally defeat Udo, to consolidate his own power at Benin and to build up the Benin Kingdom to its greatest extent.[11] Certainly trade with Europeans, tightly controlled by the *oba*, greatly increased Benin's wealth and power.[12] In the sixteenth century a few English traders came to Benin (Hakluyt 1907), followed by the Dutch in the seventeenth and eighteenth centuries (for overviews see Jones 1983; van den Boogaart 1987) and English and French explorers in the nineteenth century (Adams 1966; Gallwey 1893; Landolphe 1823; Becroft 1841). Portuguese and English accounts are short and mainly concerned with trade, whilst some of the longer and more detailed descriptions by Dutch travellers – addressing a growing European interest in the wider world – are problematic in terms of authorship, credibility and translation (Henige 1987; Jones 1987; Jones and Heintze 1987). Travellers' descriptions also reflect the changing perceptions of Africa, which became altogether darker over the course of the nineteenth century (Curtin 1965; McEwan 2000; Pratt 1992). Taking all these caveats into account, these accounts nevertheless present a valuable source in understanding Benin landscapes at the time (see 'Landscape descriptions', below).

Dapper, 1668
(Olfert Dapper did not travel to Benin himself, but collated other people's descriptions. The two versions here highlight the problem of translation.)
Jones' translation from Dutch original: 'How far the kingdom of Benin stretches from the south to the north is as yet unknown, because some places are cut off from each other by impenetrable **bush**.'(Jones 1998: 8); 'The country of Benin is very low-lying and **densely wooded**, broken up at intervals by rivers and marshes' (ibid.: 12).

My own translation from the German 1670 edition: 'How far the kingdom of Benin stretches from the south to the north is as yet unknown, because some places are cut off from each other by impenetrable **bushes**' (Dapper 1670: 486); 'The country of Benin is very low-lying and **bushy**, broken in many places and full of marshes and in others not very rich in water' (ibid.: 487).

Nyendael, 1705
(Nyendael, a Dutch merchant working for the Dutch West India Company and stationed at the Dutch Gold Coast, did travel to Benin.)
Writing about the country surrounding Gwatto, in the south of the Benin Kingdom: 'Bating the said contagion of the climate, this is a very desirable place to trade, by reason of the pleasantness of the river and adjacent country, which is very even ground, without hills, and yet rises by gentle degrees, which affords the most agreeable prospect in the world; which is yet improved by the multitude of trees which stand so regular, as if they were designedly planted in that order' (Bosman 1967: 429).
On the country around Benin City: 'the Circumjacent Country is as pleasant as could be wish'd, where no interposing Hill or Wood rudely interrupts the agreeable Prospect of thousands of charming Trees, which by their wide extended Branches full of Leaves, seem to invite Mankind to repose under their Shade' (ibid.: 466).

> **Adams, 1823**
> (A British captain who travelled widely in West Africa.)
> 'The face of the country surrounding Benin bears much of the same character as that of Ardrah and Grewhe, except that it is more thickly wooded' (Adams 1966: 110).
>
> *Adams' descriptions of the towns of Grewhe* ('which may be called the sea port of Dahomey', 6`171 north, longitd 3`6` east) *and Ardrah* (situated between Wydah and Lagos), *both in the Dahomey Gap, are as follows:*
>
> 'The country surrounding Grewhe is fertile, open, and level, exhibiting large savannahs covered with high grass, although in some parts thickly wooded with fine grown trees' (ibid.: 61).
>
> *Travelling from the coast to Ardrah:* 'the remainder of the road passes through variegated country, part of which is thickly wooded and swampy; but the greatest portion of it is open and park-like, perfectly level and interspersed with trees' (ibid.: 76). 'The surrounding country [at Ardrah] is champaign, and finely wooded' (ibid.: 77).
>
> *In contrast, he describes the kingdom of Warré, just south of Benin:* 'This country is covered with an impenetrable forest' (ibid.: 111).

Landscape descriptions seventeenth to early nineteenth century.

Landscape descriptions themselves vary, but overall give the impression of a mixed landscape with many open prospects, contrasting with the more densely wooded Warri Kingdom in the Niger Delta area. To E.W. Jones, Nyendael's description in particular suggested a 'park-like' country (Jones 1956: 104).

Descriptions and depictions of animals also point towards a more open landscape. Dapper (1668: 13) and Landolphe (1823: 15), for example, mention ostriches and hippopotami, which are both animals living in more open landscapes than forests. It is also interesting to note that a large part of the Benin ivory bought by European merchants came in the form of the long, curved tusk of savanna elephants, which Benin ivory carvings were also made from. Indeed, European traders distinguished between the different types of ivory they were buying: long and curved tusks from savanna elephants; shorter and straighter ones from forest elephants (Becroft and Jamieson 1841: 188; Ryder 1969). Moreover, there are numerous depictions in Benin bronzes, and descriptions by travellers, of horses (de Marees 1602: 236; Fawckner 1837 quoted in Roth 1968: 148; Landolphe 1823). These are often of dwarf-like horses, which are more resilient than larger horses in forested, sleeping-sickness harbouring environments, but Dapper (Jones 1998: 27) and Landolphe (1823: 35) also describe cavalry warfare in the Benin area, which again would suggest a more open environment.

European accounts also suggest that there was widespread farming. English and Portuguese visitors describe the agricultural produce they saw and bought at the Benin markets, including yam, palm wine and, in particular, palm oil; Welsh writes that there are 'great stores of palme tree' growing in the country (Hakluyt 1907: 297). Interestingly, Welsh also reports a 'great store of cotton growing' (Hakluyt 1907: 297) and Dapper that cotton grows in 'great abundance' (Jones 1998: 13); in fact, virtually all visitors to Benin mention cotton (see also Bosman 1967; Brun 1983; Landolphe 1823; Marees 1602). Cotton cloth became the most important item of export in the seventeenth century, which the Dutch bought to sell on at the Gold Coast (Ryder 1969). Even in the nineteenth century Moffat and Smith reported that 'cotton is indigenous in Benin, and is spun there and woven into cloth by women'(Becroft 1841: 192), whilst some of the elders of Iguowan in Okomu Reserve today still recall cotton being grown in the area when they were young. Cotton does not usually grow well in forest environments, so its abundance at this time could possibly suggest an altogether more open, and drier, environment.

Intriguing, in this respect, is also Nyendael's mentioning of 'great milhio': 'The Fruits of the Earth are, first, Corn, or great Milhio; for they have none of the small Sort. The large Milhio is here very cheap, but they do not esteem it; wherefore but little is sow'd, which yet yields a prodigious quantity of Grain and grows very luxuriously'(Bosman 1967: 458). As already noted by Cyril Punch, one of the earliest foresters coming to Benin in 1897, this 'great milhio' is either maize, which can be grown in cleared forest, or millet (Guinea corn), which thrives best in open country (Roth 1968: 147). Descriptions of 'great milhio' elsewhere indicate it refers to millet rather than maize (Smith 1744: 164). In view of evidence of pearl millet cultivation in southern Cameroon and southern Ghana around 3,000 years ago (Neumann 2006; D'Andrea, Klee and Casey 2001), it is perhaps not altogether implausible that millet would still have been cultivated in southern Nigeria in the seventeenth century, which would definitely suggest a drier climate and more open landscape. On balance, however, it is probably more likely to be maize, which, along with cassava, sweet potato, chilli and citrus fruits, was introduced to the area by the Portuguese in the fifteenth century. New World crops considerably diversified food crop production, with maize in particular becoming an important supplement to yam (Morgan 1959). However, it is unlikely that their introduction would have altered landscapes as radically as suggested by Morgan (ibid.) or McCann (1999), as the importance these authors attribute to the introduction of New World crops (and steel machetes) is based on the assumption that these parts of West Africa were covered in high forest and sparsely populated. Rather, maize, sweet potato and other new crops were gradually adopted and integrated into existing farming practices. Overall, traveller descriptions evoke a landscape much like the one suggested by Jones, Oates and White, consisting of a mosaic of oil palm groves, secondary forest and farmland with yam, great milhio (maize or millet), cotton and an expanding range of other crops.

It seems likely that it was in this kind of landscape that the majority of Benin *iya* earthworks were built, rather than in thick forest – in forest, the necessary clearing of trees and roots would have enormously increased an already considerable workload. *Iya* earthworks are found not only around Benin City and Udo, but throughout the heartland of the Benin Kingdom, a network estimated to be altogether 10,000 miles in length (Darling 1984). Patrick Darling, who surveyed and mapped this large network, argues that they began to be built around AD 800, as part of a second movement by Edo-speaking people from further north into the Benin area. He suggests they may have been built as spiritual boundaries, between the human world inside and the spiritual world of forests outside (ibid.). However, of the few carbon dates obtained so far, the earliest is AD 1300 in Benin City (ibid.; Connah 1975), at about the same time that, according to oral tradition, the Udo earthworks were built by Akpanigia-kon. Since the few other carbon dates currently available are all later than this, it seems possible that the majority of earthworks were built as part of the expansion of Benin administration, perhaps serving as administrative boundaries in a quite highly populated and farmed kingdom.[13]

Like Udo, many parts of the Benin kingdom retained strong regional identities and autonomy in village life and land-use practices; these will be looked at in more detail below. Nevertheless, the Benin Kingdom was highly centralised and under the ultimate control of the *oba*. As already mentioned, a number of villages and towns were ruled by *enogie*, hereditary title holders of whom many descended from junior sons of past *obas*. Other villages did not have an *onogie* and therefore somewhat more autonomy, but there was an overall system of administration of all districts (*Ikinkin Agbon-Edo*), governed by administrators known as the *onotuenyebo*, or 'he who salutes the *oba* for the village', usually holders of non-hereditary *eghaevbo* (palace) title holders (Bradbury 1957: 42; Igbafe 1980: 23). An *onotuenyebo* acted as an intermediary between the *oba* and the village headman; he transmitted the *oba*'s orders to the villages but also brought the villager's wishes to the *oba*'s notice. His main duty lay in collecting the twice-yearly taxes that all villages had to pay the *oba* – yam, palm oil, livestock and other foodstuff. To prevent the build-up of personal spheres of powers, these chiefs were not given large blocks of territory but were put in charge of scattered villages, were frequently interchanged between villages, and were also compelled to reside in Benin City, communicating with the villages only through agents. Finally, the king himself had messengers who reported to him directly, sometimes permanently stationed in the villages (Bradbury 1957: 43; see also Igbafe 1980).

Overall, the *oba*'s power rested on a highly developed system of hereditary and non-hereditary chieftaincy titles. The number of titles grew continuously, partly because of the growth of Benin's administrative power and complexity, partly because the creation of new titles was an important means for each new *oba* to secure loyal followers and to strengthen his position against existing, powerful title holders. The *eghaevbe n'ogbe*, the palace titles originally created by Ewuare, were eventually divided into three main groups: the *iwebo*, *iwegueae* and *ibiwe* palace society. Each palace society had a series of guilds (*otu*) affiliated to it, headed by one or more chiefs.[14] These fulfilled a whole range of functions: they were artisans producing wood and ivory carvings and brass works, weavers and leather workers; doctors, diviners, guardians of shrines, land purifiers and priests; dancers and acrobats; keepers of the king's harem, butchers, ceremonial farmers, and many more. All under the control of, or 'serving', the *oba*, their primary function was to attend to the needs of the *oba*, but many services, such as land purification, were also for the country as a whole. Each guild lived and worked in different parts of the city. Like Yoruba towns, Benin City is divided into different quarters (*idumu*), many of which are still named today after the guild that was originally based there.

Chieftaincy titles brought honour and status to their holders but also provided crucial access to resources in Benin's highly regulated economy. For example, only members of the trade guilds were allowed to participate in trade (Usuanlele 2005: 267). Membership of trade guilds was therefore highly sought after, especially of the guilds in charge of the lucrative coastal trade with Europeans (Igbafe 1980). In addition, administrators in the provinces received a share of the tribute they collected, whilst membership of palace societies, or of the many guilds in Benin City, also had important economic benefits.

Perhaps most important was the access to 'slaves' that chieftaincy titles provided; slaves were known as *igho n'ore uwe igho* – the real value of money (Usuanlele 2005: 273). The term 'slaves', whilst commonly used in Benin historiography, is as misleading in the Benin kingdom as in northern Nigeria; as Inikori (1996) has argued, slavery here was quite dissimilar to the slave economies of North America and rather more comparable to serfdom in Medieval Europe. Slaves were people captured during war expeditions and initially the property of the *oba*, who then distributed them amongst the chiefs involved in the campaign in which they had been captured, as a reward for their services to him (Bradbury 1957: 41, 1973d; Igbafe 1980). Alternatively, chiefs could purchase slaves at 'hinterland' markets, a common investment for chiefs who had accrued wealth through fief holding, control of political patronage or long-distance trade (Bradbury 1973c). Chiefs with many slaves founded 'slave villages', in which slaves cleared land and farmed for their owners. They were also allowed to farm for themselves, and many such slave villages eventually developed into normal villages, with the same social organisation as other villages (Bradbury 1957: 45). In addition to whole slave settlements, wealthy chiefs could also make arrangements with the headman (*odionwere*) of a village to station their farm workers there and to farm some land (ibid.). Overall, the status of these captives was not fundamentally dissimilar to that of other dependents of 'big men' – indeed, all Edo men were regarded as the slaves of the *oba*, even if a clear distinction was made between these and real slaves (Bradbury 1973a: 133, 1973d: 181). Benin slavery was symptomatic of a society in which control over labour was vital to prosperity (Guyer and Belinga 1995), but Benin did not participate in the transatlantic slave trade (Ryder 1969).

The fact that chieftaincy titles were so instrumental for anyone wishing to build up wealth and power meant that these non-hereditary titles considerably bolstered the *oba*'s powers, as he alone could award titles. In theory, titles were supposed to be given to particularly worthy citizens, who had in some ways earned the award of a title. In practice, however, they often went to those who could afford to buy them – the sale of chieftaincy titles was in fact an important source of income to the *oba* (Bradbury 1973b: 80; Igbafe 1980: 22).

Patrimonialism was therefore deeply engrained in the Benin political and economic system: ultimate power rested in the person of the *oba* and the distribution of resources was determined by patrimonial power structures, namely chieftaincy titles and political allegiances. Benin's patrimonial political and economic organisation did not disappear in the colonial period; despite some initial political upheavals, the existing elite was in the best position to exploit the new opportunities offered by political and economic changes. This context, as will be explored in the following chapters, shaped the implementation and outcomes of forest policies throughout the twentieth century.

However, it is important to stress that patrimonial power was never all encompassing in the Benin Kingdom. As Bradbury carefully demonstrates, there were various mechanisms to check the *oba*'s and his title holders' powers. The power of the *oba* himself was curtailed by the *uzama*, whose hereditary status gave them more independence than other title holders (Bradbury 1973d). In the country, villagers had some means to stop abuses of power by the *enogie* through the *onotueyebvbo*, who gave them direct access to the *oba* (Igbafe 1980). The village, with its social structures based on gerontocracy rather than patrimonialism, was a much older political unit than the Benin Kingdom itself, and villages had considerable independence in the running of their affairs (Bradbury 1973a, 1973d). Rural life and land-use patterns will be looked at more closely within the next section on Udo and Iyek'Ovia.

Village, Farm and Forest in Iyek'Ovia

Even though many Udo inhabitants had fled after its defeat by Esigie, the town retained its regional role as the capital of Iyek'Ovia district, the part of the Benin Kingdom west of the Ovia river. Whilst some town quarters were abandoned altogether, others remained; in the early colonial period there were five quarters, each headed by a different title holder (Marshall 1939). Since Oba Esigie's decree Udo's overall ruler was the *Iyase N'Udo* who was appointed by the *oba*, but new *iyases* tended to come from the same Udo family. All the villages of Iyek'Ovia were 'under' Udo, which meant that they paid tribute to Udo rather than to Benin and were generally administered by Udo. Many villages had also been deserted after Udo's defeat in the early sixteenth century, but several remained and, over the course of time, new settlements were founded.

Of the villages found in the area today, the oldest are Igueze and Urezen, west of Udo, which trace their foundation to people who emigrated from Benin during the time of Oba Ewuare.[15] Others, like Ugbelegan, Iguafole, Utesi and Iguowan, all just south of Udo, were founded by people from Udo in more recent times. In contrast, north of Udo, Iguoshodi and Iguobazuwa – the headquarters of Ovia South-West LGA today – were originally slave villages, founded by Benin chiefs (Igbafe 1980: 24). In the southern parts of today's Okomu Reserve, Ijaw-speaking fisher people settled in villages on river banks, but it is unclear when this took place; today both Edo- and Ijaw-speaking people claim they were the first to arrive in the area.

Unlike Ijaw settlements, Edo villages were always founded away from river banks, in order to avoid tsetse flies and other river-based harbingers of diseases. River banks would have been more thickly forested and more difficult to clear, and the soil along rivers was of poorer quality for farming. This was already observed by Nyendael in the eighteenth century: 'The Soil, a little distant from the River, is extraordinary fruitful; and whatever is planted or sowed there, grows very well, and yields a rich crop. But close to the River the Land is not good; for tho' what is sown comes up, yet contagious Damps of the River kill it' (Bosman 1967: 459). Still today cassava and plantain grown in the grey, clayey soil near rivers do poorly, and people generally avoid planting there if they can help it.

The foundation of a new Edo village involved the *ogiefa*, one of the most important Benin chiefs, whose role was to purify and 'appease' the land (*oto*). For a new place to become a true 'village', a place distinct from the bush (*oha*) or forest (*egbo*) around it and one in which you can have intercourse, an *ikhimwin* tree (*Newbouldia laevis*) had to be planted, a tree that is said never to die (Melzian 1937). Every Edo village has one or a small group of such founding trees (*inyator*), and its size can give some indication of the age of a village.[16] The founding ceremony had to be performed by the *ogiefa*, involving the sacrifice of a cock or goat and prayers. Whilst the *ogiefa* of Benin fulfilled this role for most Edo areas, Udo had its own *ogiefa*, the *Ogiefa N'Udo*, who was responsible for Udo and the villages under Udo.

Through the pivotal role of the *ogiefa* in its foundation, each new Edo village was from the beginning incorporated into the Benin Kingdom, with those in Iyek'Ovia under the more direct rule of Udo. As described above, all villages also had government representatives and paid taxes to the *oba*, although those in Iyek'Ovia paid tribute to Udo, rather than Benin City itself. But Edo villages also had considerable autonomy in their own government; as Bradbury emphasised, the village was the basic political unit of the Benin Kingdom (Bradbury 1973d). Village government rested primarily on gerontocratic rule by elders and on a male age grade system similar to others throughout West Africa. Boys were members of the *iroghae*, responsible for clearing paths and other light duties. Young adult men constituted the *ighele*, who performed heavier and more skilled tasks, including military service. Older men were members of the *edion* (the elders), who formed the village council. This village head was the *odionwere*, the oldest man of the village. Boys were initiated into the *iroghae* upon reaching puberty, but the transition into subsequent grades was more fluid and depended as much on achievement and public esteem as on age itself; the *odionwere* was therefore not necessarily actually the oldest man. Nevertheless, overall the *edion*'s authority in villages rested in age, and the qualities of wisdom, experience and knowledge of traditions associated with it. As Bradbury has argued, these principles of village government counterbalanced as well as complemented the patrimonial structures of the kingdom's administration as a whole and helped to limit the *oba*'s and chiefs' powers (ibid.).

One important area in which villages had control over their own affairs was land. Officially, the *oba* was the owner of all Benin land; this was expressed in the decree 'oba o re yan oto' (Bradbury 1973d: 132). The *oba*'s ownership played a role when non-Edo people wanted to fish or collect palm produce within the kingdom. They had to liaise with the *asuen* (title holder), on behalf of the *oba*, in order to gain his permission (Usuanlele 2005; von Hellermann and Usuanlele 2009). But local people did not need the *oba*'s direct permission to gain access to farmland; rather, farmland was allocated by local communities themselves. Villagers who wished to start a new farm could choose their own plots, which then had to be approved by the village elders, headed by the *odionwere*. Decisions about the location of a new farm depended, in part, on a farmer's access to labour; the clearing of high forest required more work than that of recently farmed fallow land. Overall, the larger and most important farms were located far away from villages, with a few smaller farms or kitchen gardens closer by (Bradbury 1957). In Udo, for example, big farms were always outside the moat, a considerable distance away. This was in order to protect farms from domestic animals like goats and dwarf cattle kept in the villages, but also because people believed farmland far away, deep in the forest, to be particularly good for yam.

Areas of particularly fertile soil were identified by first assessing the vegetation, which gave some indication of soil quality, and then by testing the soil itself. I was told in Iguowan that 'in the old days, in the high forest, they will go to a certain area, dig far into the ground, smell the soil, then you say no good. Then you go to another place, dig again, meet mud soil, smell, then say yes, here will be good'.[17] Good soil for farming, *oto ugbo*, was a mixture of *oto ekhia* (sandy soil) and *oto ulakpa* (mud soil).[18] *Eken ne uloka* (loose soil), was also preferred, because sticky and clayey soils could cause the rot of seed yams (Usuanlele 2003). The choosing of land took place in the dry period in January or early February. Men then cleared prospective farms at the end of the dry period, in February and March. They cut down most small trees but left those that were too large to cut, such as the silk cotton tree (*Ceiba pentandra*) and those that were useful to preserve. The most important of these was the oil palm, of which every part was used in different ways – for cooking oil, kernels, palm wine, brushes, leaves for roofing, and other purposes (Collins 1945; Zeven 1967, 1972). Another tree left was the iroko (*Milicia excelsa*), the royal tree. Iroko could only be cut by the *owina ne igbesamwan*, the woodworkers guild working for the *oba* (von Hellermann and Usuanlele 2009). In addition, some young trees were left standing to be used as climbers for yam tendrils and as shade trees.[19]

Cut-down wood was left to dry for a few days and then everything was burnt, with some trees protected against fire. In April, the first crop to be planted was yam, the most important staple crop. Planting consisted of putting seed yam, kept aside from last year's harvest, in the soil at regular intervals. Yam was considered a man's crop, although women assisted in weeding and planting. Women were also responsible for all other crops; after a man had cleared his farm he divided it into different plots between his wives or other women in his family. They planted plantain, maize, okra, cocoyam, melon, beans, pumpkin, tomatoes and chilli pepper in between the yam, in such a way so as to minimise the spread of disease and competition between different roots for water and minerals. Women were responsible for providing food for their husbands and children, but could sell or trade any surplus to their advantage (Bradbury 1957: 23–24). In May additional yam poles, branches kept from the trees cut down before, were planted in the ground, in order to give support to the growing vines of the yam, which needed to be carefully arranged around the poles and the ropes between them. There are many different kinds of yam – white, yellow and water yam in particular, and these mature at different times, between September and November.

After harvesting the crops – with yams stored in specially built barns in villages – farms were left to lie fallow for a number of years, although plantain could still be harvested for another year or two. The length of the fallow period varied; in some cases it may have been fifteen to twenty years, or longer, with some farms abandoned altogether, but many plots were farmed again after about eight years. It seems likely that there was shorter rotation on the farms closer to the villages, where crops other than yam were planted, and longer fallow periods for the yam farms further away. On an abandoned farm, bushes and trees soon started to grow. According to early-twentieth-century observations in nearby Oluwa forest (Allison 1941), the first taller tree to take over from shrubs after just a few years would have been the rapidly growing *Musanga smithii*, the umbrella tree; indeed, this is still the most common tree on fallow land today. After eight years or so *Musanga* still dominated at Oluwa, but numerous other light-demanding species, including *Ficus asperifolia*, *Riconondendron africanum* and *Albizzia spp*, appeared. After about twenty years these were the tallest trees, but there were also common under-storey rainforest species such as *Diospyros spp*, whilst the mass of the upper storey was formed by light-demanding timber species like *Guarea cedrata*, *Triplochiton* (obeche) and *Celtis soyaruxii*. Overall, farming resulted in a definite increase in quick-growing timber species, including also *Mansonia* and *Entandrophragmas* (sapele wood), all of which, as light demanders, grew far better on abandoned farmland than in high forest (ibid.). In young secondary forest in Okomu itself, Jones observed many saplings of large emergent species, such as *Khaya ivorensis* (mahogany or lagos wood), *Irvingia gabonensis*, *Alstonia boonei*, *Antiaris africana*, *Terminalia superba* (white and black afara) and *Lophira alata* (ironwood) (Jones 1955: 574). *Milicia excelsa* (iroko) too often grows up on abandoned farms, as well as many other rainforest trees that thrive best in open or partially open conditions in the early stages of their lives (Hawthorne 1995). Through shifting cultivation, then, the inhabitants of Iyek'Ovia and other parts of the Benin Kingdom fundamentally shaped the landscape and vegetation around them. All settlements were surrounded by a mosaic of farms, fallow land and secondary forest at different stages of growth, all containing fast-growing timber species that thrived on abandoned farms. Further away from villages, where farms were sparser or where, as in Okomu, large areas of farmland had been abandoned altogether after conflict and migration, high forest grew up, still containing many of the emergent species that had started to grow after farming.

Within villages themselves a variety of shade and fruit trees were planted, including avocado, guava, orange and the indigenous *otien* fruit tree (*Chrysophyllum africanum*). Villages were often surrounded by more fruit and shade trees, in particular kola nut, which also grew along the paths leading into a village and between villages and farms (Fig. 2). They were planted or grew from accidentally dropped seeds. Today the sites of former Edo settlements, like those elsewhere in West Africa (Fairhead and Leach 1996), can still be easily identified by a high concentration of fruit and shade trees as well as their founding *ikhimwin* tree, the *inyator*, even if they have become part of a larger forest. In many of today's inhabited villages too, the same fruit and shade trees still grow along paths, and the *inyator* remains a central feature of the village (Fig. 3). But perhaps the most prominent vegetation features of villages, still today, are sacred groves containing shrines to ancestors. Some of these are quite small and mainly consist of coconut palm trees – coconuts play an important role in many Edo rituals. Visible from afar, the sight of coconut groves instantly signals a settlement to an approaching visitor. Bigger groves on the outskirts of villages contain a diverse range and often very tall timber trees; such tree clumps, too, indicate a nearby settlement. Sacred groves form an integral part of the Benin landscape, as indeed in many other parts of Africa (Sheridan and Nyamweru 2007).

Figure 2 Path leading to Igueze village, lined by kola nut and fruit trees. Photograph taken by the author, 7 November 2002.

Figure 3 The *inyator* of Igueze village, a group of *ikhimwin (Newbouldia laevis)* trees planted at the foundation of the village. Photograph taken by the author, 7 November 2002.

This ecologically variegated landscape, then, also had different social uses and meanings for its inhabitants. In the immediate surroundings of villages there were many areas of fallow land and secondary forest between village and high forest that were accessible to all people, with no spiritual association. All Edo inhabitants utilised a large number of plants for medicine, timber and other uses, and many had in-depth botanical knowledge (Darling 1995; Hide 1943; Melzian 1937). But high forest, far away from villages, was thought of as dangerous and 'spiritual'. Before venturing into such forests it was necessary to fortify oneself spiritually, and only hunters, priests or 'doctors' – *osun* (leaf) specialists – were able to do this. People believed that forests were inhabited by various powerful spirit beings (*erinmwin egbo* and *oso*) and dwarfish spirits (*eseeku*), while the iroko trees were the supposed lairs of witches' (*azen*) covens (von Hellermann and Usuanlele 2009). Thus, the further a native doctor had to go and find a particular leaf, the more potent it was considered to be, as he could share some of the powers of the witches through his knowledge of the forest. No one visited the forest during *eken* (the fifth day of the week), because this day was reserved for the spiritual beings to utilise the forest; violation could bring disaster on the person or community (ibid.).

Moreover, many rivers, streams and lakes, as well as hills and depressions, were believed to be powerful ancestors who had turned themselves into landscape features. These places became shrines or sacred groves, which could only be entered by priests or chiefs on special occasions, just as shrines within villages themselves were looked after by chiefs. In Udo, the Odighi lake where Aruaran drowned has since then been a sacred place (Fig. 4). Even more important in the region as a whole was the Ovia river and the Ovia cult associated with it. Ovia was a legendary historical beauty who had married the King of Oyo and who had melted into a river when his jealous older wives plotted against her (Bradbury 1973c). There were several Ovia shrines in Iyek'Ovia, of which the one at Urezen was and remains of particular importance.

The Edo landscape was thus as rich in meaning and social uses as in ecological diversity, an ecological diversity that was created through human land-use patterns. When forest policies such as forest reservation and logging control were introduced over the course of the twentieth century, these existing practices were fundamentally changed, but also in turn shaped the outcomes of adopted policies. The ecology of the forests of Benin was the product of a particular history, which significantly affected foresters' attempts to manage high forests. The last stage of this history was the nineteenth century, when political turmoil and the kingdom's decline brought with it further forest growth.

Figure 4 Odighi lake near Udo, where legend tells us that Aruaran drowned. Photograph taken by the author, 6 November 2002.

The Decline and Fall of Benin: Forest Growth

If there were conflicts and upheavals throughout Benin history, these began to increase from the end of the seventeenth century onwards; Nyendael, coming to Benin in the early eighteenth century, already found it in a ruinous state (Ben-Amos and Thornton 2001). Unwilling to loosen its tight state control over trade with Europeans, the Benin Kingdom found itself increasingly isolated. By the late nineteenth century European merchants conducted more and more trade with traders in the south rather than with Benin rulers, and economic and political conditions in the kingdom deteriorated. As the *obas* reacted to political unrest with ever more autocratic rule, many people fled and population levels dropped throughout the Edo areas. Nineteenth-century depopulation was a widespread phenomenon in West Africa, which was caused by epidemics such as small pox as well as warfare (Fairhead and Leach 1998). Disease may also have ravaged the Edo population, as Jones (1956) speculates.

A comparison of nineteenth-century descriptions of the route from Gwatto to Benin City, the main route that all visitors to Benin City took (see below), suggests that the political turmoil and depopulation of the late nineteenth and early twentieth centuries may have led to forest increase.

> **Adams, 1823**
> 'The course of the road from Gatto to the capital is about NE by N, and the road passes over a country nearly level, intersected with deep woods and swamps' (Adams 1966: 110).
>
> **Moffat and Smith, 1841**
> 'The next day they were carried in their cots to the city of Benin. Distant from Gatto about 20 miles, in a north-easterly direction, the country on their route being finely wooded, and in some places very beautiful' (Becroft 1841: 190).
>
> **Punch, 1889**
> 'The road after leaving Egwatu is a mere bush path and during the wet season is converted into a running stream'. Later along the road, he mentions a 'two hours' march through dense forests' (Punch 1889: 18–19).
>
> **Gallwey, 1893**
> 'Benin City lies about 25 miles from Gwato, the whole route being through dense forest, with the exception of the last mile or two' (Gallwey 1893: 128).
>
> **Boisragon, 1897**
> *Describing the typical West African bush they were walking through on their way to Benin:* 'if one tries to imagine a thick wood in which big and little trees all intermingle their branches, with a tremendous dense undergrowth of shrubbery of all sorts, with brambles and various other evildoing thorns, all woven together into a maze so thick that neither man nor beast can press through it, one comes somewhere near the idea' (Boisragon 1897: 94).

Descriptions of forest on route from Gwatto to Benin City in the nineteenth century.

These descriptions convey a change from a more mixed and open landscape to one thickly forested throughout the entire journey from Gwatto to Benin City. However, this evidence still needs to be treated cautiously. For one, it may reflect changes in attitudes towards West Africa as much as actual landscape changes, as the more benign portrayals of West African peoples and landscapes of seventeenth- and eighteenth-century visitors gave way to increasingly lurid descriptions of the dangers, evils and 'dense forests' of West Africa in the late nineteenth century (Pratt 1992); Boisragon's account certainly is part of this 'adventures in the African bush' genre. Moreover, the impressions gained from walking only along this path may also have been misleading. This was pointed out by H.N. Thompson, the first Director of Nigeria's Forestry Department, after his first survey of the forests of Southern Nigeria at the beginning of the twentieth century:

> [E]xtensive forests like those met with in the moist zone of Burma, through which the forester can wander for days without coming across the traces of

human habitation, are practically absent here. It is true that persons may by travelling merely along the native roads and paths come to an opposite conclusion, but they have only to strike out to the right and left of such lines for a short distance to find that the usual belts of forest left along roadsides quickly gives place to farm lands. (Thompson 1906: 261)

This suggests that forests had not completely taken over by the early twentieth century; there was still a large amount of used or only recently abandoned farmland, and most of the forest that was there, as Thompson continues, 'consist[ed] of secondary growth that ha[d] sprung up since the native farms were last abandoned' (ibid.). Nevertheless, if not everywhere, some forests did grow as a consequence of depopulation and warfare in the late Benin Kingdom and may well have continued to expand in the early twentieth century. J.F. Redhead, who worked as a forester in the area in the 1950s and 1960s, describes how in the course of his work he came across many *iya* earthworks in forested areas. Discussing the history of the Benin Kingdom as a whole and its gradual decline from the seventeenth century onwards, Redhead speculates:

It is likely that, when the Benin Kingdom went into decline, settlements now marked only by ditch and mound earthworks, began to be abandoned. Existing mature forest, within or near these earthworks, is not likely to be older than the beginning of this period ... Historical accounts and tradition, and the presence of many ditches and mounds throughout the forests of Benin Division, point to the former existence of a very large population. (Redhead 1992: 116–17)

Nineteenth-century depopulation due to disease, warfare and slave trade had similar effects elsewhere: the ancient Yoruba town of Oyo was destroyed in the late eighteenth century, which led to the growth of high forest in previously more savanna-like conditions (Allison 1962; Keay 1947), and Fairhead and Leach (1998) describe large-scale reforestation, in the basis of traveller reports, in Sierra Leone, southern Ghana and Liberia. The canopies of the majority of forests throughout West Africa were therefore dominated by tree species that do not regenerate in forest conditions (Aubréville 1938; Letouzey 1968; Poorter et al. 1996; Richards 1952), largely because virtually all forested areas of the early twentieth century had been inhabited or farmed at some point in their history (Hawthorne 1996; van Gemerden et al. 2003). Importantly, these tree species included almost all of the timber species that were to become the focus of colonial scientific forestry. However, even though foresters such as Thompson, Redhead and others were clearly interested in and understood the dynamic history of the Benin forests – which had created an abundance of desirable timber species – the policies they adopted in order to manage these forests, as we shall see in subsequent chapters, took only limited account of their history.[20]

Benin's history took a tragic turn in 1897, when Benin City was captured by a British 'punitive expedition', launched in retaliation for the 'Benin massacre' (Boisragon 1897), at which a party of British traders who had approached Benin City despite warnings that they were not welcome (it was the time of a Benin festival), had been attacked and some killed. Following the capture of Benin City and its incorporation into the Niger Coast Protectorate in 1897, the *oba*, Oba Ovoramwen, refused to agree to the arrangements offered to him by the British and was sent into exile in Calabar. With this, the existing political structures, which the early colonial officials were also quite ignorant of, became temporarily redundant; instead, in this interregnum period, a Native Council was founded, in which various individuals were appointed as paramount chiefs over large territories (Igbafe 1979). The then *Iyase N'Udo* was close to Ovoramwen and was supposed to be sent into exile with him, but committed suicide before the British had captured him. Consequently, the army was sent to Udo and took away Udo's bronzes and antiquities (Ezele 2002b); Udo had become part of the British Empire.

Notes

1. Shaw himself was cautious about interpreting these as indicators of cereal production, since he assumed that the area was forested at this time (after an open period around 8000 BP), but in view of the large amount of data now available pointing towards the dry period beginning around 4000 BP, it seems they can now be more positively associated with cereal production. This is corroborated by evidence of pearl millet agriculture at this time from southern Cameroon and southern Ghana, considerably further south than millet agriculture today (D'Andrea, Klee and Casey 2001; Neumann 2006).
2. The Edo language belongs to the Niger-Congo language family.
3. In view of the findings in Cameroon (Neumann 2006) and the interpretation of Iwo Eleru proposed here, perhaps in the Benin area, too, there was cereal as well as yam and palm oil production.
4. This is the standard account, as related by Egharevba (1968) and others. In recent decades a second version has emerged, according to which it was in fact the last *ogiso*'s banished son who had become the King of Ife. These diverging accounts reflect different historical claims: in the standard one Benin claims greatness for Oranmiyan and subsequent *obas* through their direct descent from Ife, the oldest and most important Yoruba Kingdom; in the second, Benin tries to claim its superiority over Ife.
5. The Oranmiyan story is today hotly disputed; there are now various alternative versions trying to prove that the banished son of the last Ogiso, Ekhaderan, had in fact become the king of Ife and that Oranmiyan was his son. These different versions are part of ongoing attempts of Benin historians to portray Benin as both connected to Yoruba states yet also independent.
6. The *uzama* were a fusion of existing chiefs (known as *edion*, the elders, during the *ogiso* period) and some of the chiefs accompanying Oranmiyan from Ife (Egharevba 1968; Bradbury 1973c).

7. There are fierce contestations of this at Udo: Here, many claim that in fact both the street names of Udo and its chieftaincy titles were all copied by Edo (Benin). See Imafidon (1987: 140).
8. There are competing accounts of the events leading up to Aruaran's death: according to Egharevba, Aruaran drowned himself to avoid capture by Benin. According to Udo oral history, Aruaran won his last battle but due to a misunderstanding his servant thought he had lost and threw all of his belongings in the lake; Aruaran jumped in after them and drowned.
9. Osemwughe and his followers fled westwards, but pursued by the *oba*'s forces, eventually surrendered, promising to pay tribute to the *oba* from now on from his new home in the west. The people with him were called *Emwa n'Udo* (the Udo deserters), which later became Ondo. Ondo is a large Yoruba city today, and Ondo State neighbours Edo State in the west today. Osemwughe was mispronounced Osemawe, the title by which the rulers of Ondo have since been known (Egharevba 1968: 27). Osemwughe and his followers were not the only people to leave after the destruction of their city. According to Ezele, some went to the Ishan area where they founded Udo, whilst others founded Obogo near Aladja, Madaja, and Udu now known as Udu clan in Urhobo land near Warri (Ezele 2002b: 7; Okpewko 1998: 8). Some also went to Onitsha. There have been speculations in fact that Onitsha's legendary founder, Eze Chima, may have come from Udo; the Onitsha shrine where kings are consecrated, at any rate, is known as Udo (Meek 1937: 12; Ohadike 1992).
10. Ezele (2002b: 9) adds: 'he was also given the power to bestow titles to good, noble and responsible citizens in the name of the Oba of Benin.'
11. The role of Portuguese firearms in Benin's rise to power is disputed (Graham 1965), but there is much evidence for it. For a fuller discussion of this, see von Hellermann (2005).
12. At this time the main items of trade were pepper, ivory and slaves, all of which the Oba had a monopoly over (Ben-Amos and Thornton 2001). The trade was well organised, with some palace chiefs, the *uwangue* and the *eriba*, members of the *iwebo* palace association, being placed in charge of the waterside trade (Igbafe 1980: 31) The *iwebo* palace society, which today looks after the king's regalia, was created by Oba Ewuare with the specific purpose of looking after the newly acquired coral and red cloth. Although at the time of Oba Ewuare the Portuguese probably did not reach Benin City itself but only Gwatto, the importance of European trade was already great then (Bradbury 1973e: 34).
13. Other possible explanations for the building of the moats include protection against elephants (Clutton-Brock 1999) and water drainage (Connah 1975).
14 According to Marshall's 1939 Intelligence Report of Benin City, there were then sixty-eight of these guilds. For a complete list see Igbafe (1979: 392–395).
15. I was told two different accounts of the foundation of Urezen. In both of these the defeat of the people-eating monster *ihebezu* by a man called Ize featured prominently (interviews with Urezen inhabitants, 15 January 2003).
16. *Ikhimwin* trees are also planted by individuals on their compound, for similar founding purposes. They are also often used for washing lines.
17. Interview with Alfred Ohenhen, Iguowan, 4 February 2002.
18. Ibid.
19. The National Archives (NA), CO 879/88, Enclosure No. 84, Report on the Agricultural and Forest products of the Central and Eastern provinces of Southern Nigeria, by Gerald C. Dudgeon, 27 June 1906. See also Hawthorne (1995).
20. For a fuller discussion of these contradictions between foresters' observations and knowledge and the policies adopted, see von Hellermann (2011).

CHAPTER 2

Separating Farm and Forest
Reservation and Dereservation

Introduction

At 460 square miles, Okomu Reserve was once one of the largest forest reserves in southern Nigeria. Today, however, it has been converted to a range of different uses. Two large expatriate-managed oil palm and rubber plantations take up most of the northwest and northeast of the reserve, and an oil palm and food-crop plantation has been set up in the north by a local businessman. There are also numerous small-scale cocoa and plantain plantations, whilst food-crop farming under the Taungya system is widespread throughout the reserve. Only the heart of the reserve, a national park since 1999, is now fully dedicated to forest protection (see Map 1). The conversion of reserve land to plantations and farms is not restricted to Okomu: Edo State's once extensive reserves, which in 1937 covered 64 per cent of the then Benin Division,[1] were reduced to 21 per cent of land by 2002 (Forestry Department 2002a: 435) and were more recently estimated to amount to only 12 per cent.[2] What remains also often contains little forest, as more and more reserve land is converted for other uses.

Encroachment into reserve land and the conversion of reserved forests for agricultural uses is widely regarded as one of the major causes of forest destruction in Edo State (Forestry Department 2002a). To many observers the lack of protection of reserves presents one of the most visible manifestations of the post-colonial state's failure to manage its forests properly. Conservationists, forest officers and local farmers alike believe that forests were well protected in the colonial period, and that it is mainly due to Nigeria's political and economic decline in recent decades that reserves have ceased to function (Oates 1999: 125–26). Like elsewhere (Klopp 2000), they have become a 'source of patronage' for the ruling elite, who freely allocates reserve land to political cronies (Lowe 2000).

If broadly accurate in outline, this general understanding of the unfolding of reservation in Edo State is nevertheless incomplete, even misleading. Rather than presenting an abrupt departure from well-functioning forest protection in earlier parts of the century, dereservation and forest conversion in recent decades are part of a longer continuum of political and economic processes and in many ways directly linked to reservation itself. Moreover, just as dereservation today has a range of environmental outcomes, not all of which are equally destructive, so reservation did not necessarily have the ecological effects foresters hoped for. This chapter is concerned with the full story of reservation and dereservation, with the continuities between them and with the actual ways they have transformed both landscapes and livelihoods in Edo State over the course of the twentieth century. In this way it presents not so much a counter-narrative but an important corrective to dominant accounts, teasing out the continuities and complexities in the relations between reservation and dereservation.

This chapter, then, begins by looking at reservation itself. If colonial forest reservation in the Benin Division is remembered in forestry circles today as a purely technical measure, integral to rational forest management and successfully implemented, historical investigation reveals it to have been rather different. From the beginning, reservation was intensely political and played a central role in emerging land conflicts. After a slow start in the early twentieth century, by the 1930s reservation became a key strategy for the Benin *oba* to maintain control over land, in the face of growing private land acquisition for oil palm and in particular rubber plantations. There was, by then, substantial resistance to further reservation, not just from elite planters but also rural farmers, whose access to land was increasingly curtailed by both. Even though their protests did not perhaps amount to large-scale resistance movements such as described elsewhere (Guha 1989; Peluso 1992), they signalled the growing rivalry between different forms of land use and the tensions reservation was already creating and experiencing. Furthermore, reservation not only fundamentally disrupted the symbiotic farm and forest relations that, over the centuries, had produced the timber-rich Benin forests, it also, by putting such large tracts of land under government control, created the legal conditions and structures of land control that underlay subsequent dereservation and forest conversion.

It is against this background that the dynamics of subsequent dereservation need to be understood. Importantly, substantial dereservation had already begun in the late 1930s, as soon as the last wave of reservation had been completed and long before Nigeria's political and economic crisis of the 1980s and 1990s. Competition for land between agriculture and forestry grew in the late colonial period and independent Nigeria in the 1960s, with agricultural production becoming an increasing government priority. Large-scale dereservation only resumed in the late 1970s, but pressure from agricultural interest groups had been building up for many years. Moreover, it was because reserves were under government control – unlike the unreserved land surrounding them, which was tightly controlled by communities – that reserves were now essentially more easily available for plantation projects than community land and indeed became a source of patronage for politicians. On a smaller scale too, Okomu Reserve became a source of patronage for local communities who allowed Yoruba cocoa farmers to settle in it, whilst food and plantain farming also expanded. Largely 'illegal', small-scale farming in the forest reserve nevertheless addressed real needs for agricultural land, even more than larger-scale projects. Indeed, the illegal expansion of small-scale cash crop production within reserves can be regarded as policy change from below, through informal processes rather than official changes.

With regard to the social outcomes of forest conversion, social as well as environmental, again distinctions need to be made between large-scale, government-endorsed rubber and oil palm plantations and small-scale cocoa and plantain farms. Both provided income opportunities and enabled more rapid local development than would have been possible without the opening up of Okomu Reserve, especially in Udo. But whilst the cultivation and marketing of plantain and cocoa was largely peaceful, improving the livelihoods of Udo market women as well as farmers, the larger stakes that the expatriate-managed plantations involved contributed to a political crisis that gripped Udo in 2006. Moreover, the recent expansion of larger plantations has endangered many small-scale plantain, cocoa and food producers, who have been forced to leave Okomu. Small farmers have therefore ultimately been more vulnerable and at risk, even if they contributed to arguably more sustained local development than the larger plantations. Similarly, the environmental effects of forest conversion, whilst generally less destructive than standard accounts suggest, have varied significantly between small- and larger scale plantations. Cocoa farmers have frequently been described as particularly destructive, but both cocoa and plantain farming in Okomu Reserve preserves far more original tree cover and biodiversity than the monocultures of the large oil palm and rubber plantations. The nuances of forest conversion's social and environmental impact will be explored in detail in the final section of this chapter.

Reservation: Dividing Farm and Forest

Forest reservation was the cornerstone of scientific forest management. It was essential to gain exclusive control over forest tracts in order to accurately calculate the forest estate and to write working plans regulating logging and regeneration activities. As H.N. Thompson, the Chief Conservator of Forests of Southern Nigeria, explained to the Liverpool Chamber of Commerce in 1904: the main stages of 'forest organisation' begin with 'the determination of [the forest's] limits and its legal status; that is to say, the area in question must be demarcated, surveyed; the existence of all rights contained in it settled, and its working prescribed in a plan called its "working plan"'.[3] Without exclusive control over a forested area, scientific forestry was impossible to practice. This, however, could not be and was never a sufficient rationalisation for reservation. Instead, the main justification given by colonial foresters was that reservation presented the only means of effective forest protection. The main threat to forests, in Nigeria as elsewhere in the tropical world, was perceived to be shifting cultivation, widely regarded in colonial circles as a highly destructive, 'pernicious system of cultivation' (McLeod 1908: 242). Throughout the colonial world there were frequent alarmist reports about the destruction caused by shifting cultivation, which could only be halted through reservation (R. Bryant 1994; Chandran 1998). Ralph Moor, familiar with the discourses against shifting cultivation then current in West Africa and also concerned about forest destruction through rubber harvesting, saw reservation as amongst his first tasks as the High Commissioner of the newly created Niger Coast Protectorate (Omosini 1978). However, urgent as it was deemed to be for forest protection and management, forest reservation had a slow start in the Protectorate.

Beginnings 1901–1914

The main reason for this slow start was that the creation of forest reserves presented certain legal challenges to the colonial government. In theory, the former Benin Kingdom had conquered territory status, which meant that the British Crown had ultimate control over its land (Elias 1951). But with the exception of Crown land assigned for urban development projects (Igbafe 1979), in practice the colonial government did not insist on this control and did not use it as the legal basis for reservation. In contrast to northern Nigeria, where all land was indeed declared Crown land, in the south the overall policy was to respect and uphold 'customary' land tenure,[4] and local authorities were recognised as the 'owners of the land'. Their consent was necessary for the creation of reserves, as it was important to the colonial government to follow such procedures in order to give reservation a veneer of legality and popular consent. It wished to avoid both criticism in Britain and popular protest within the Protectorate. High Commissioner Moor learnt this lesson with the first attempt at forest reservation, the 1901 Southern Nigeria Forestry Proclamation. Due to the Protectorate's 'conquered territory' status the Proclamation's wording ('The High Commissioner may from time to time constitute any waste or forest land at the disposal of the Government a Forest Reserve') seemed to suggest that virtually the whole territory of the Protectorate would be reserved, which caused widespread outrage amongst local communities and British commercial interest groups, as well as the London-based Aborigines Protection Society (Egboh 1978: 83). The Proclamation's generally highly ambiguous wording meant that, in practice, no reserves could be created under it at all.

After several further unsuccessful attempts (see Egboh 1985), the Forest Ordinance of 1908 finally provided the necessary legal framework for forest reservation.[5] This made provisions that local communities had to be notified and their consent sought before any land was declared a reserve, and that the rights and privileges of the communities over their lands and forests had to be respected (ibid.: 47). These rights and privileges referred to hunting and fishing only, not farming. On the whole, reserves were to be under the control of the government, or the Forest Department, but managed on behalf of local communities, who were still recognised as the official 'owners'. It was under the 1908 Ordinance that Okomu Reserve was created, along with two others (Jamieson and Ologbo) in the then Benin District. During his first tour of the Benin and Sapele Districts, in 1903, Thompson had identified the 'upper valley of the Osse [Ovia] River' as one of the richer forest tracts of the Benin District.[6] In 1907 it was inspected 'with the view to the establishment of Forest Reserves'(McLeod 1908: 241), and in 1910, 'the consent of the owners to reservation' of 300 square miles was obtained.[7] On 17 April 1912 it was constituted a reserve.[8] These 300 square miles form the southern part of today's Okomu Reserve. It is the high forest that grew in the aftermath of the defeat of Udo by Benin City in the early sixteenth century.

With Oba Ovaramwen in exile in Calabar, Udo and Iyek'Ovia were then, like the rest of the Benin area, ruled by a paramount chief appointed by the colonial government. Whilst no documentation is available today, it is likely that it was this paramount chief, Chief Obahiagon, together with Udo chiefs, who gave his consent to the creation of the reserve, since both paramount chiefs and local villages were then officially recognised as 'owners' of forests (Igbafe 1979: 266). Given Udo's predominance in the area, it seems unlikely that the smaller Edo villages south of Udo, or indeed the Ijaw fishing villages along the Ovia and Okomu rivers, were consulted at all.

In the absence of any detailed documentation, it is also difficult to establish how and why Chief Obahiagon and the Udo chiefs agreed to forest reservation. They may have been motivated by the prospect of increasing royalty payments from timber concessionaires, which were equally divided between paramount chiefs and local villages (ibid.). The area reserved then was also far away from Udo and the Edo settlements surrounding it, and reservation would not have affected them directly. In general, the earliest forest reserves in Southern Nigeria were far away from settlements. Like Okomu, they typically constituted 'buffer zones' between communities; 'land that has purposely been left intact ... between belligerent tribes,' (Thompson 1911b: 131). Moreover, Okomu and other reserves initially existed really more on paper than on the ground, with no demarcation of the reserve border and only minimal control from the small numbers of forest guards employed then. It seems likely that early reservation therefore had relatively little impact on people's livelihoods, which may explain why there are no records of resistance to reservation at this time. For the same reasons, it would also have had relatively little immediate ecological impact, as it did not change land-use patterns in the area.

In these early years there was also as yet relatively little pressure for land from tree-crop cultivation, namely oil palm and rubber. Palm oil had been exported since the mid nineteenth century and – 'greasing' industrial development in Europe – was, by the early twentieth century, by far the most important export from Southern Nigeria (Martin 1997). The eastern provinces of the Protectorate of Southern Nigeria, with their extensive palm groves, were historically most associated with palm oil production (Martin 1988), but in the Benin area, too, a considerable amount was produced. However, as throughout West Africa, at this time palm oil derived entirely from wild or semi-domesticated oil palms (Zeven 1967). In the Benin District the collection of the fruits and the preparation of oil and kernels for the market were mainly done by Sobo- (Urhobo) speaking people. These were generally engaged by a headman, who had obtained permission from the (Edo) villagers to collect produce from communal land, and were paid on a commission basis.[9] The colonial government did not interfere with these production methods, for in contrast to other colonial powers in West and Central Africa, who encouraged the establishment of large expatriate-owned plantations, the policy adopted by Moor, Egerton (Governor of the Lagos Colony) and other governors in British West Africa, was to encourage local production. Moor in particular advocated this policy, which fitted his overall vision of local development in Southern Nigeria and complemented the policy of retaining traditional land tenure systems (Afigbo 1970). It also made sense politically and economically, as the development of peasant commodity production required less effort and social outlay (Freund 1998: 111; Phillips 1989). For all these reasons, the colonial government in Southern Nigeria actively curtailed concessionaire plantations and refused the repeated applications William H. Lever (later Lord Leverhulme) made to acquire land for oil palm cultivation (Phillips 1989; Udo 1965).[10] Palm oil export therefore did not lead to any significant alterations of existing land-use practices.

Rubber was the second key tree crop in the area. Like palm oil, rubber was an important product in the development of Europe's industries, and the late nineteenth and early twentieth century saw the great 'rubber boom'. The rubber forests of the Benin Kingdom, described by several of the visitors who came to the city in the 1880s and 1890s, had been amongst the key incentives for its capture in 1897 (Punch 1889; see also Anene 1966: 144). For Ralph Moor, who had witnessed their destruction in the Lagos Colony, concern over the protection of rubber trees was a prime motivation for the setting up of the Forest Department in the then Niger Coast Protectorate, and rules regulating rubber tapping were issued even before logging rules.[11] In contrast to palm oil, rubber production soon switched from the harvesting of wild indigenous rubber trees (*Funtumia elastica*) to plantation production, where *Funtumia* as well as a few imported Brazilian *Para* seedlings were planted. This was organised through the Forest Department, which at this time combined both agriculture and forestry. It initiated and supervised the establishment of local rubber plantations in Edo villages, and, by 1910, there were almost a thousand plantations in Benin City and adjoining districts (Thompson 1911a: 12). In fact, it was along the Benin–Siluko road, i.e., in the area around Udo and particularly in Udo itself, that rubber plantations were first started and soon became established village features (Vetch 1912). These plantations were under the charge of a headman (*oga*) who was exempted from other kinds of work, but all farmers had to provide labour for the plantations (ibid.). In addition, the Forest Department encouraged chiefs to establish plantations, using forced labour from the villages under their jurisdiction (Igbafe 1979).[12] The department was pleased with the popularity and rapid spread of rubber plantations. It spent a considerable amount of time on research into best planting and tapping methods, on the preparation of coagulated rubber into 'biscuit rubber', and on training local people in these methods (Vetch 1912). These efforts came to an end in 1910 when Thompson was able to 'realise his long-cherished ambition' to separate forestry and agriculture and to concentrate the efforts of the Forest Department on the creation of the permanent Forest Estate (Nicolson 1969: 96; Thompson 1911a). Under the aegis of the Agriculture Department, rubber cultivation continued but with less enthusiastic support from the government once the rubber boom came to an end and prices began to fall, which had already happened by the outbreak of the First World War. Overall, with generally low population levels and both reservation and rubber plantations only just beginning, there were at this time no recorded conflicts over land between villagers, plantation owners and the Forest Department. This began to change with the gradual expansion of both tree-crop production and forest reservation in the following decades, leading to first conflicts in the 1930s.

From the 1916 Forest Ordinance to the Benin Forest Scheme

After the amalgamation of Northern and Southern Nigeria in 1914, a first comprensive Forest Ordinance for the whole of Nigeria appeared in 1916, which laid the foundations of Nigerian forestry for many years to come. It was drafted by Thompson at a time when, due to the First World War, the creation of a regular timber supply from the colonies was a British priority (Worboys 1979). The ordinance made systematic forest reservation a central aim. Following the example of other colonies, it stipulated that 25 per cent of Nigerian land needed to be reserved in order to meet present and future timber needs (Lugard 1970: 435). Taking advantage of the emergency situation of the war, the ordinance even contained a clause that enabled the compulsory constitution of forest reserves (Lugard 1970: 439; Omosini 1978: 69), but in practice reservation continued to depend on the cooperation of local authorities. For this reason the ordinance also contained the promise that restrictions on local forest use would be lifted, once its aim of 25 per cent of reservation had been achieved. These restrictions, also introduced by the 1916 Ordinance, forbade the felling of a vast range of tree species and were highly irksome for local people, essentially criminalising their normal farming activities. In order to make the incentive of the promised removal of forest-use restrictions more persuasive, in 1921 the Forest Department entered separate agreements with southern Nigerian chiefs, whereby the reservation of 25 per cent of land was now to apply to regions, rather than the whole of Nigeria (Egboh 1985: 69).

In the meantime the amalgamation of Nigeria and the introduction of indirect rule to southern Nigeria – indirect rule, whereby colonial government is largely administered by local rulers, had been used successfully by Lord Lugard in Northern Nigeria and was therefore now also established in the south – coincided with Oba Ovoramwen's death in exile. The colonial government decided to restore the Benin monarchy, and Oba Ovoranmwen's son was installed as Oba Eweka II (Igbafe 1979). This was fortuitous for the purposes of forest reservation, as now the *oba* was once again the official 'owner of the land' throughout Benin Division, and all forest reservation could legitimately be negotiated with him alone. If not a complete 'invention of tradition' (Ranger 1983), in the context of reservation the nature of the *oba*'s control over Benin land became far more direct than it had been in the pre-colonial period. Moreover, the *oba*'s overall position – both vis-à-vis the colonial government and vis-à-vis the paramount chiefs who were unwilling to give up the powers they had enjoyed during the interregnum – was weak to begin with and made it easier to lean on him to agree to reservation. At the same time, the restrictions on local forest use, which had been continuously reviewed to increase the number of protected species of trees and were strictly enforced with prosecutions,[13] were vastly unpopular in the Benin Division. There were widespread protests against the ordinance and its strict implementation, adding further pressure on Oba Eweka to agree to forest reservation (von Hellermann and Usuanlele 2009). By 1928 four new forest reserves – Obaretin, Gilli-Gilli,[14] Ohosu and Owan – had been created, bringing the total land reserved to 1,214 square miles, almost 30 per cent of the land area (Ainslie 1930). The locations of these new reserves, all on the outer borders of the Benin Division, suggest that Eweka may well also have seen strategic advantages in reservation. They covered contested areas where several neighbouring groups were settling – Ohosu Reserve bordering Yoruba-speaking areas in the west and Obaretin and Gilli-Gilli reserves in the south, where Urhobos and Itsekeris were moving in. As the borders of the Benin Division had already been redrawn several times following a series of contestations, reservation presented a way of consolidating Benin – and the *oba*'s – hold over land (von Hellermann and Usuanlele 2009).

With almost 30 per cent of land reserved in the Benin Division, the target of 25 per cent was already well exceeded by 1928. However, because the Benin Division was the most important timber area in southern Nigeria, the Forest Department had no intention of stopping reservation and justified this by continuing to rail against the evils of shifting cultivation. For example, Thompson's successor, Ainslie, wrote in the annual report of 1928: 'the rapid extension of reserves is the only way in which this menace can be held in check ... it is a race between the conservative activities of the Forest Department and the destructive activities of the shifting cultivator'(Ainslie 1929: 7). The extension of Okomu Reserve by a further 160 square miles in the north was the next reservation project envisioned in the Benin Division. Here, however, Eweka put in significant resistance, because even though the target of 25 per cent had already been exceeded, the restrictions on local forest use had still not been lifted. He refused to sign the agreement for the extension until this was finally done in June 1929 – albeit excluding all existing concession areas, which, much to everyone's frustration, made the lift virtually meaningless.[15]

In the early 1930s, triggered in part by the Dust Bowl experience in the United States, there was a renewed wave of concern amongst colonial administrators in Africa over the destructive effects of shifting cultivation and over environmental degradation and desertification (Anderson 1984; Beinart 1984; Fairhead and Leach 1998; Swift 1996). These concerns led to the organisation of two extensive tours of Nigeria by two external experts, E.P. Stebbing in the north and Major J.N. Oliphant in the south. In their wake, in particular due to Stebbing's alarmist reports over the spread of the Sahara (Cline-Cole 2000; Milligan and Binns 2007; Stebbing 1937), there were many calls for further reservation.[16] In 1933 the 1921 shift to applying the 25 per cent rule to regions was declared non-valid, which at a stroke made it possible to press for further reservation in some areas, including, of course, the Benin Division.[17] During his tour of southern Nigeria, Major

Oliphant had evaluated the 'commercial possibilities' of its forests,[18] and it was decided to set up a pilot scheme for intense forest management. The area chosen for this was the Benin Division, still the most important timber area of Nigeria. Moreover, the Benin Division was again politically ideally suited for such a scheme, as Oliphant noted in his report: 'It has an enlightened ruler in the *oba*, who has full control over the land.'[19]

This *oba* was Akenzua II, who had succeeded his father after Eweka's death in 1933. Educated at King's College in Lagos, Oba Akenzua II had some training in the Native Authority of Abeokuta, and was generally deemed broadly supportive of colonial government. He was indeed a useful negotiation partner; on 1 April 1935 he agreed to the Benin Native Administration Forest Scheme, and thereby to the reservation of a further 1,000 square miles of land. Given his education and his perception of himself as a modern ruler, he may well have been persuaded by foresters' arguments for the need to preserve forests for future generations. In his memoirs, the forester St. Barbe Baker (1942) describes how he preached passionately to the *oba* about the dangers of deforestation, which, St. Barbe Baker claims, made a deep impression on Akenzua. At the same time, for Akenzua as for his father, reservation presented a useful way of maintaining and establishing control over land. This concerned not only neighbouring groups and other outsiders, but also a growing number of chiefs and other individuals claiming land for private plantations, as will be explored in more detail below.

Perhaps his most important motivation was that the Forest Department had decided that both old and new reserves would come under the Benin Native Administration (BNA). The idea of involving local authorities directly in forest regulation and reservation had already been used in the early struggles with reservation in Lagos Colony (Grove 1997), but it was the 1927 Forest Ordinance that first empowered local authorities to constitute Native Administration Reserves (Egboh 1985: 71). Any royalties, licence fees and other revenue from these reserves would go to the Native Administrations, rather than to the Nigerian Forest Department. In return, the Native Administrations were responsible for the wages of forest guards and the salaries of the forest officers dealing with the allocation of licences. Under the Benin Forest Scheme, as a 'valuable experiment', the whole of Benin forestry was put under the BNA; since 'half measures and any degree of dual control', as Oliphant argued, would be undesirable, this included the handing over of existing government reserves.[20] It promised substantial financial benefits in the form of timber revenues and royalties to the BNA, which would increase the *oba*'s overall power and ability to govern.

If the *oba* and his council were persuaded by such incentives to give their consent to the scheme, and thereby to the reservation of a further

1,000 square miles, the proposals were met with fierce protests from plantation owners, who saw the expansion of commercial rubber and oil palm plantations threatened by further large-scale reservation. At this time the expansion of plantations was actively promoted by the Agriculture Department. The rather relaxed attitude of colonial governments in British West Africa towards local agricultural production had begun to change since the late 1920s, when more interest began to be taken in its development. This was part of a broader shift in colonial policy towards greater investment in the development of the colonies, which culminated in the Colonial Development and Welfare Act of 1940. Because of growing competition from Malaysia and the Congo, palm oil production, in particular, was increasingly encouraged in West Africa (Phillips 1989), now with active government promotion of oil palm plantations. In 1928 a programme was started with 21 acres of African-owned plantations, which had risen to 4,172 acres by 1936. Sixty-three per cent of these plantations were in Benin Province – in fact, mostly in the Benin Division (Ekundare 1973). According to the 1936 annual report of the Agriculture Department, 'continuous propaganda and the example of well-kept Native Administration and School plots [were] steadily breaking down opposition to planting and the idea of cultivating palms [was] now coming to be regarded with less suspicion and scorn in the more backward areas'.[21]

With rubber prices still low in the 1920s, rubber production was far less actively promoted than oil palm production. But rubber prices began to recover slowly in the 1930s, and rubber cultivation too expanded considerably at this time.[22] Many of the new oil palm and rubber plantations were set up by urban title holders, rich traders and the educated elite, who began to invest their money in plantation cash crops. They felt threatened by the proposal of further large-scale forest reservation and, with the support of the Agriculture Department, started a campaign against the Benin Forest Scheme. They wrote a petition to the Governor, protesting against the new reserves and demanding the reservation of land for plantations.[23] This demand was resisted by Akenzua, who argued that anyone wishing to expand an existing plantation had to get permission from the Native Authorities, explaining that:

> The object and reasons for looking for this safe-guard is to be sure that the remaining Benin land does not unceremoniously fall into the hands of private families. As the Reserves are being constituted for the benefit of the Benin people – present and future generations – there seems to be no reason why private families should have exclusive right to make use of the remaining land outside the Reserves unless and until they can show that their rights so to do are derived from the proper source i.e., if they acquire land in the usual way by applying for them from the Oba in whom Benin lands are vested from time immemorial.[24]

56 | *Things Fall Apart?*

The *oba*'s apprehensions about the expansion of private plantations emerge clearly from this passage. Indeed, there was reason for concern over the ways in which urban title holders obtained land for their plantations. They increasingly converted their existing food-crop farms in villages to cash-crop plantations and extended them annually by using the labour of their peasant relatives and others whom they employed to look after the plantations in their absence.[25] Such land appropriation through the establishment of private cash-crop plantations became widespread (Fenske 2011; Igbafe 1979: 311), and there were many complaints from rural people about community land being taken over by chiefs and the urban elite.[26] Receiving such complaints, Oba Akenzua II himself became increasingly concerned about land alienation, which affected not only local people but also his own overall hold on land. He wrote repeatedly to the district officer and others to avoid the alienation of land.[27] In this context, as discussed above, the creation and protection of reserves presented to him a way of maintaining control over land and played a role in his support of the Benin Forest Scheme.

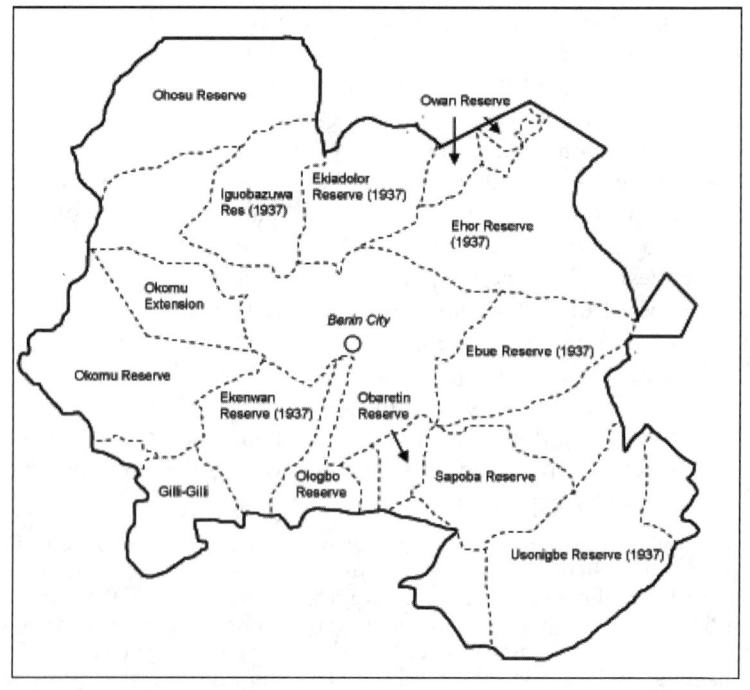

Map 4 Existing and new reserves demarcated under the Benin Forest Scheme in 1937. The areas indicated do not show enclaves already included in the 1937 demarcations, which is why the reserved area seems larger than the 64% of land it covered. This is adapted from a map of Benin Province drawn and reproduced by the Nigerian Surveys Department in 1933, corrected in December 1934, which showed the new reserves as proposals. NNAI, BP 999.

Whilst urban planters maintained their campaign against the Forest Scheme, their protest initially had little impact. By 1937 an additional 1,377 square miles were reserved, through the creation of Okomu Extension, Iguobazuwa, Ekiadolor, Ehor, Ebue, Usonigbe and Ekenwan reserves. This brought the total land under reservation to 2,631 square miles, or 64 per cent of the Benin Division. In fact, overall, reserves covered an even larger area – 80 per cent of land (see Map 4) – but all new reserves had so-called 'enclaves' carved out for local communities, which brought down the total slightly. Such vast additional reservation made the Benin Division the most extensively protected division of Nigeria and presented a considerable victory of the Forest Department and the *oba* over the Agriculture Department, urban planters and rural populations. This victory, however, was relatively short lived. It soon became obvious that all these new reserves infringed on or included village settlements and farmland. Already in 1936 the Senior Assistant Conservator of Forest wrote that 'a number of villages and small camps in these reserves have complained of shortage of farmland and have been allowed to farm for this year only pending investigation of their claims'.[28] In 1937 Chief Conservator Weir noted that 'it is evident from the preliminary examination that the proportion of non-productive land within [the new reserves] is much higher than was anticipated in earlier estimates'(Weir 1938: 2). Discoveries of substantial amounts of farmland within the reserves were repeatedly reported in subsequent years (Oliphant 1941: 2).[29] Unsurprisingly, reservation everywhere now met with far more protest from rural populations as well as planters, which had some effect. In fact, dereservation started almost immediately after the creation of the new reserves. Before turning to this, however, both the local and the larger ecological and political implications of this last wave of large-scale reservation will be considered by looking in detail at Okomu Extension.

Okomu Extension

As mentioned above, Okomu Extension was proposed as early as 1927 under Oba Eweka II, but it was not until 1935 that it was officially constituted a reserve. In part, the delay was caused by Eweka's death in 1933 and his succession by his son Akenzua II. But the main reasons were problems in determining the exact boundaries of the reserve and the enclaves within it. In contrast to the original Okomu Reserve, its extension directly affected Udo and many of its surrounding villages, whose inhabitants strongly objected to their farmland being taken away or drastically reduced in size. The determination of the reserve boundaries and of the amount of enclave land allocated to different communities now engulfed by the reserve lay with the Reserve Settlement Officer (RSO), appointed for the task of 'judging' the correct boundaries. The RSO in charge of establishing the boundaries of Okomu Reserve appears to have been conscientious and scrupulous, as the thick file on the lengthy negotiations over boundaries with local communities documents.[30] But his attempts to be rational and fair were, in many ways, rather misguided.

The initial enclaves suggested by the RSO resulted in complaints from residents of a large number of villages – Udo, Ikoha, Etete, Ugolo, Urezen, Iguolaho, Iguokahen, Gbelegan, Utesi, Iguafole, Iguowan and Igue-Agbado. After a tour of the complaining villages, the RSO reported:

> With the exception of Igue-Agbado I visited all the dissatisfied places myself, and as far as possible inspected both the enclaves and any alternative sites that the people wished to advance in preference to those which had been allotted to them. At the risk of incurring criticism myself, I cannot refrain from remarking that due regard has not been paid (in apportioning the land) to any ratio btw. the farming population and the land itself. In connexion with this I cite as an instance the village of Iguafole which has been given an enclave of 1sq mi, although the number of farmers is only 29, whereas the combined villages of Ikoha, Etete and Ugolo with a farming population of 25, 52 and 87 respectively (164) have only received 1.6 sq mi between them.[31]

Following this first investigation negotiations continued for several years, and some Okomu communities at least were successful in obtaining larger enclaves than they had originally been allocated. The Ikoha, Etete and Ugolo enclave increased from 1.6 square miles in 1930 to 5 square miles in 1934, and that of Udo and Iguafole from 2 to 5.3 square miles. Iguowan was less successful: its enclave merely increased from 1 to 1.1 square miles.[32] Finally, all sides came to an agreement, and on 2 May 1935 the Okomu Extension, an area of 160 square miles, was constituted a government reserve.

At Okomu, as at other reserves, the RSO may have tried to provide adequate farmland for communities in enclaves according to the number of male adult inhabitants, and many communities indeed successfully negotiated larger enclaves for themselves. But while the idea of establishing a 'fixed ratio' in determining the appropriate size of enclaves may have been well intended, it forced a radical departure in land-use practices, in a number of ways. It did not, for example, allow for any spiritual meaning and uses of land and the RSO at Okomu had little time for the claims of the people at Urezen, still today famous for its powerful Ovia shrines:

> The objections of the Urezen people are religious ones, there being several Ovia jujus in the land which has been assigned to them. I am inclined to think that their claim for land on the right hand side of the path, to the east, is a merely perverse one and that if they are dissatisfied with the land given them they should go and farm outside the reserve to the north of the old Udo–Siluko road. Certain of them have taken out permits to farm within the reserve and I recommend that they should be allowed to continue to farm within the reserve but on the land allotted to them. It will then be seen whether the religious plea is a frivolous one or not.[33]

Furthermore, the 'fixed ratio' idea did not account for existing land-use practices. As described in the last chapter, people usually had the most important farms, their yam farms, far away from settlements; only smaller garden farms, including cassava ones, were in the immediate vicinity. When the reserve was created, people were forced to abandon their farms inside the bush and had to concentrate farming within their enclaves or on community land outside the reserve, radically shifting their farming practices.

It is not surprising that reservation therefore inevitably featured as the most central watershed in historical accounts of the region in the interviews I conducted with elders. In an interview with the chief of Iguowan, for example, I noted:

> In those days, we Bini, we believe somehow, once you have your family, everything is free, anywhere you move to, automatically that place becomes your settlement. Nobody will come to disturb you there. Then: the time of the reserves. Enclaves.[34]

In Udo, too, older farmers recall how they had to abandon many of their farms when Okomu Extension was created:

> When the reserve was created, people didn't like it at all, having their land taken away from them. But because it was government, they could not do anything. They had been farming inside the reserve before, but were asked to pack and leave, compartment by compartment.[35] The vegetation inside then gradually reverted to forest.[36]

Even if the RSO and the Forest Department as a whole may have sought to protect local communities' farmland, and if through petitions and protests some local communities were able to change the outline of the reserve boundaries, it emerges clearly from these memories that reservation nevertheless detrimentally affected people's access to land and livelihoods. If before they had farmed in different plots scattered widely around villages – cassava and vegetable farms on more recently farmed fallow land nearby, yam farms cleared deeper inside the forest – people were now forced to abandon their farms in the forest and to farm in much closer proximity to their villages and to other farms. This division of all land into separate spheres of forest and farmland is central to a European vision of forest management and indeed of landscapes, but was quite alien in southern Nigeria, where landscapes had resulted from centuries of symbiotic interrelations between farms and forests. Furthermore, whilst the fundamental disruption of these patterns through reservation greatly reduced people's access to land, it did not actually present the effective measure of forest management it was supposed to.

Reservation did, of course, protect larger forest tracts from disturbances other than foresters' own planned silvicultural treatment programmes, exactly as scientific forestry demanded. But it had little impact on overall tree cover, since farmers now simply had to farm elsewhere, clearing land of trees outside rather than inside reserves. Biodiversity overall probably decreased, even if species composition inside forests gradually changed towards less light-demanding ones. The clearing of forest for farming, followed by a fallow period, does not reduce, in fact can increase, biodiversity (Guyer and Richards 1996). Outside reserves a gradual reduction of fallow periods due to greater pressure on land reduced opportunities for a diverse range of trees to regenerate. Most importantly, reservation did not actually serve the central aims of forestry at the time, namely the regeneration of economic timber species. As discussed in the previous chapter, the composition of Okomu and other Benin forests was shaped by centuries of farming. Many of the trees that dominated the canopy of mature forests did not naturally regenerate in forest conditions, including the timber species foresters were interested in protecting and fostering – 'mahogany', or lagos wood (*Khaya ivorensis*), iroko (*Milicia excelsa*), sapele wood (*Entandrophragma cylindrical*), obeche (*Triplochiton scleroxylon*), black and white afara (*Terminalia superba*) – all secondary species that needed light in the early stages of growth. Whilst protecting forests as a whole, reservation therefore actually curtailed the regeneration of timber species inside forests.

Foresters were, in fact, quite aware that economic species were not regenerating well inside high forests. As early as 1904 Thompson noted:

> [T]he constitution of the stock, so far as the African "mahogany", "cedars" and "walnuts" are concerned, is very unsatisfactory and typical of what usually occurs in an unexploited tropical forest. Instead of there being a regular gradation of plants from the seedlings upwards to the exploitable size, we have seedlings, suppressed saplings and, comparatively speaking, a large number of trees of exploitable size, the intermediate gradation being practically absent.[37]

Timber regeneration within forests remained a constant problem, as we shall see in the next chapter. But by forcing local farmers to concentrate their farming on community land outside reserves and shortening fallow periods here, reservation also made the regeneration of economic species *outside* reserves more difficult. Whilst these effects would perhaps not have been immediate, reservation nevertheless fundamentally disrupted the conditions that had fostered the growth of economic species in West Africa in the first place, by creating separate spheres of forest and farm.

At the same time, reservation drastically changed land ownership in the Benin Division. Whilst it is true that in the pre-colonial period the *oba* had been the nominal owner of the land, communities still had control over the land surrounding them and could freely choose where to farm. Reservation removed more than half of Benin land from communities and put it under government control; it meant that the *oba* became more literally 'the owner of the land' than he had been before. Placing reserves under the BNA in the 1930s might have had the appearance of local control, and in some ways may be seen as a precursor to community conservation efforts today (Wardell and Lund 2006). However, the BNA was far removed from actual local communities, and at the point of independence reserves came under state government control, confirming the alienation of forests from local communities that, in reality, had already taken place before. These radical changes in control over forest lands came to matter in the context of rising competition over land through the emergence and spread of cash-crop production.

Farm or Forest? Dereservation

Initial Dereservation in the 1930s and 1940s

Whilst the Forest Department and Oba Akenzua initially managed to proceed with reservation despite resistance from plantation owners and local farmers, this victory did not last for long; no sooner had reserves been created, in 1937, than investigations into dereservation began in all those places which, as mentioned above, included large amounts of land already farmed. At the same time, urban planters maintained their campaign against the reserves through petitions and newspaper articles.[38] This was all the more the case after the introduction of the Permanent Crops Order of 1937, designed to curtail land appropriation through the establishment of oil palm and particularly rubber plantations (Igbafe 1979). The order was worked out at a meeting of the *oba*, the Assistant Chief Conservator of Forest and the District Officer in response to the ongoing complaints of peasants about the shortage of land resulting from permanent crops cultivation. It was strongly supported by the *oba*[39] but highly unpopular amongst the chiefs and rich urban dwellers affected. Combined with other, equally unpopular measures imposed by the Native Administration, it culminated in a mass protest meeting in December 1940, following which the Native Administration was reorganised to include a wider spectrum of chiefs and even 'ordinary' citizens (ibid.: 360), giving the planting lobby more power than before.

The outbreak of the Second World War meant that palm oil and now also rubber production needed to be stepped up significantly to meet British wartime needs. The numerous measures taken by the Agricultural Department to increase palm oil production included the distribution of oil presses and competitions organised by missionaries and the *oba* in 1943.[40] Meanwhile the loss of Malaya to the Japanese in 1942 meant that British authorities now also encouraged rubber production in African colonies (Fenske 2010). Due to government propaganda but also rising prices, a great number of new rubber plantations were started in the Benin Division. Observing these developments with some worry, the District Officer wrote to the *oba* in 1943 to warn him that 'if Benin people continue on their present road there will in a few years be a serious shortage of farmland and a large excess of rubber trees, which in all probability will be valueless.'[41] The government was particularly worried as this expansion followed the same patterns as before:

> The device by which uneducated farmers were induced, in many cases by city-dwellers, to sow rubber seeds in their farms must be condemned: in return for a small outlay on his part the absentee "landlord" becomes the owner of the rubber trees and thereby gains a long lease over the land. The net effect of this and of villagers planting for themselves is that much farm land has gone over to cash crop small-holdings.[42]

The expansion of oil palm production followed similar patterns (Igbafe 1979: 311) and food crops too were used by some urban chiefs to claim land for themselves in this way, causing complaints from village communities.[43] These processes presented a second transformation in rural land tenure concurring with forest reservation, from community control under which community land was freely available to all its members, to effectively private land ownership, under which the sons of urban elites, because of the permanent rights vested in tree crops, could inherit tree crop plantations directly from their fathers (Fenske 2010; von Hellermann and Usuanlele 2009). Overall, there was increasing competition for land between village communities and urban planters. In this context, encroachment into reserve land by farmers increased. This had already occasionally happened for some time: in 1926, for example, Ijaw farmers were found to be farming in the south of Okomu Reserve (Thompson 1927). But from the late 1930s a growing number of illicit rubber plantations were set up inside reserves, including in Okomu Extension (Mutch 1952: 13).[44] In fact, so many of the new reserves were found to be containing areas of farmland with little forest that they proved increasingly impossible to maintain, and first dereservation measures had already begun by the late 1930s. By 1943 substantial parts of the new reserves had been dereserved: over 100 square miles from Iguobazuwa Reserve, 226 square miles from Ebue Reserve, 98 square miles from Ekiadolor and 300 square miles from Usonigbe Reserve.[45] With ongoing protests and petitions by local people,[46] further dereservation was carried out in Ehor Reserve, reducing its original 400 square miles to 115 square miles in 1944.[47] By 1948 total reserved land was calculated to be 39 per cent of the Benin Division, down from 64 per cent in 1937.[48] Thus, substantial dereservation had already started long before the political and economic decline of the last few decades.

'Agriculture Must Take Priority over Forestry': 1940s to 1970s

After the end of the Second World War the Colonial Development Act was put into full effect and spending on the development of the colonies increased considerably (Young 1988). In Nigeria the first series of ten-year development plans were published, including a Ten Year Plan for Forest Development. It was drawn up by the Chief Conservator F.S. Collier, following general guidelines from the Governor of Nigeria, Sir Arthur Richards. The plan signals a shift in the colonial government's priorities away from forestry and towards agricultural development: it begins by stating that the main object of forest policy is 'the production of the maximum benefit to the greatest number from the minimum amount of forest which is essential for the general well-being of the country' and further states that 'agriculture must take priority over forestry' (Collier 1948: 5).[49]

This policy made it increasingly difficult for the Forest Department to hold onto and protect its forest estate. Initially, however, having already conceded substantial dereservation in the Benin Division, the Forest Department continued to insist on the need for forest reserves and to defend its forest estate. In 'Policy – How Much Forest is Necessary?', D.R. Rosevear, Chief Conservator in the early 1950s, wrote that, as 'the chances of securing any further appreciable addition to the estate in the high forest zone' were 'almost nil, the best we can do – and the least from a long-term administrative view-point – is to hold on to what we have got' (Rosevear 1952c: 3–4). Initially the Forest Department largely succeeded in doing so. There was increasing pressure for dereservation from people wishing to establish large-scale plantations, with some success (Falola 2004), but on the whole the government of the then Western Region resisted most suggestions of dereservation, as Chief Conservator Burgess noted with some pride in 1956 (Burgess 1956: 2). The following year further proposals were made for dereservation, but the Forest Department again managed to persuade the government against this decision (Burgess 1957). On the whole there were only minor changes in the extent of the forest estate in the 1950s.[50]

After independence, with the First National Development Plan of 1962–1968, the Nigerian government launched a programme of encouraging further large-scale plantation agriculture through quite substantial investment and incentives. Considering the long history of resistance to expatriate-owned plantations and a strong preference for peasant agriculture under colonial rule, it was somewhat ironic that the government now sought to attract foreign investments in plantations (Udo 1965). Again, there was considerable pressure to use forest reserves for these new plantations. This pressure arose in part because of land shortages outside reserves – there was increasingly little spare land outside reserves, most of which by now was farmed or occupied. But it also arose because of the difficulty obtaining community land for large-scale plantation projects. When urban chiefs had appropriated community land in the 1930s, this was largely done through existing ties to villages and was still relatively small scale. From the 1950s onwards, in contrast, it was the regional agricultural development boards who were seeking to establish large-scale plantations on good land (Falola 2004). For these, as well as for any potential foreign investors, it was more difficult to obtain large areas of land from communities. Free of community claims, unoccupied and unfarmed, forest reserves presented an attractive alternative. The Nigerian geographer Reuben Udo, for example, recommended their usage in 1965: 'The forest reserves … offer considerable stretches of unsettled land, most of which would be suitable for plantations. These reserves are already under the control of government so that the problem of local land tenure, which features in land negotiations with the natives, will not arise' (Udo 1965: 361). This passage highlights not only that forest reserves were fully under the control of the government – by then all notions of 'on behalf of the people' had long disappeared – but also that this meant that reserve land was potentially easier to obtain access to than community land.

A few large, commercial, foreign-financed plantations were established, which, by the mid 1960s, were responsible for 8 per cent of cash-crop production (Helleiner 1966). However, possibly because the Forestry Department[51] continued to defend its reserves successfully, still no immediate large-scale conversion of reserves to rubber or oil palm plantations ensued; instead, some plantation projects managed after all to obtain community land. These included the first large-scale plantation that was established in the Okomu area, the then government-owned Osse River Rubber Estates Ltd (ORREL). This plantation was created on unreserved community land just outside Okomu Reserve. Like elsewhere, the project initially met with resistance from people in Udo who did not want to give up their land, much of which at the time was still covered with small-scale rubber plantations. However, Chief *Oliha N'Udo* saw the plantation as an opportunity for development and, he recalls, persuaded other Udo chiefs to accept the plantation. One reason why Udo chiefs and elders in the end decided to give their consent may well have been that it was not Udo's own farmland that was affected but that of Ikpopa and Etete, two small farming villages just northeast of Okomu Reserve and under the control of Udo. People living here probably had little say in the matter. 1780 hectares of land were granted to the plantation, and the farmers affected had to find other places to farm (see Map 3). This illustrates some circumstances under which community land could be gained for plantations, namely when more powerful communities have control over large areas, including other villages.

However, in general community resistance remained fierce, and the establishment of new plantations was rather slow. The Biafran Civil War (1967–1971) disrupted the drive towards the expansion of the plantation economy and reduced the production and exports of palm oil and rubber as a whole. In the 1970s, however, the Nigerian government and its advisers became increasingly concerned about low agricultural productivity, especially as the transformation of the economy through oil was expected to increase demand (Kirk-Greene and Rimmer 1981). Palm oil production, which had fallen considerably, was a particular concern. With the Third National Development Plan 1975–1980, the military regime of General Murtala Mohammed tried to revive and expand Nigeria's flagging oil palm industry. Again, however, the difficulty of obtaining rights in adequately sized land presented major restraints.

These difficulties were supposed to be addressed by the Land Use Decree of 1978, which placed all of a state's land in the hands of the Governor of the state (Osemwota 1989: 75). Ostensibly, the Land Use Decree was meant to be 'for the use and common benefit of all Nigerians', but as Paul Francis points out, this rhetoric of equality and justice only shrouded its real intended use – as an instrument of expropriation (Francis 1984: 5). However, Francis' prediction that 'the majority of the peasantry are likely to escape the benevolence of a "rationalisation" of their land tenure system' (ibid.) has been largely correct in Edo State. Some communities, it is true, have seen their land sold by their leaders to large-scale developers, in particular those villages under the control of Benin chiefs rather than village elders themselves. For example, the chief of Iyekeze, a village in Ovia North-East LGA near the Udo–Benin road, sold all community land to an Isoko man, who established an oil palm plantation there. These are again circumstances similar to those of Udo, where local residents themselves have comparatively little power vis-à-vis their leaders. However, such plantations on community land are relatively rare, and on the whole the decree has had little impact in rural areas.

With ongoing communal stronghold on land tenure outside reserves, reserve land remained not only the most attractive for large-scale plantations, it was also more easily accessible precisely because it was under government and not community control. In fact, whilst the Land Use Decree had little effect on community land, it may well have reinforced the state government's sense of actual ownership of forest reserves. With the decree forest reserves came under the direct control of state governments, and the Forestry Department's power over reserves was reduced. According to the *Forest Resource Situation Assessment of Nigeria*, one result of the Land Use Decree was that 'private investors should not have difficulties in obtaining land for forestry development'(Olaleye and Ameh 1999). The decree may therefore ultimately have had rather more impact on reserve land than on the community land it was targeting. Indeed, it was in the late 1970s that the latest wave of dereservation began. The programme of large-scale government oil palm plantations outlined in the Third National Development Plan would not have been possible to implement without access to reserve land. By the early 1990s medium- and small-scale farming, too, began to take off within reserves, much of which was allocated through patronage networks. These different processes will now be looked at in detail with regard to Okomu Reserve. But as I have tried to show here, the increasing prioritisation of agriculture over forestry since the 1940s meant that pressure for further dereservation had been building up for several decades. Dereservation resulted from broader shifts in policy and the need for agricultural expansion as much as from weakening control of the Forestry Department over forest reserves and patronage politics.

Dereservation and Forest Conversion in Okomu Reserve 1960s to 2000s

As part of Murtala Mohammed's effort to revive Nigeria's palm oil industry, the Nigerian government commissioned Socfinco, a Belgian firm of agricultural management consultants, to undertake a study of the rehabilitation possibilities for oil palm cultivation in Nigeria in January 1976 (Egbon 1990). Okomu Reserve was amongst four areas identified as suitable for the establishment of a large-scale plantation.[52] In the same year the Okomu Oil Palm Company (OOPC) was set up as a federal government project, taking over an area of 16,000 hectares in the western part of the reserve (see Map 3, Introduction) on the basis of a ninety-nine-year lease by the state government. The creation of OOPC, then, followed an official federal government decision to use reserve land for these plantations, taken in the course of a nationwide attempt to foster agricultural production. It did not result in the immediate conversion of its entire area from forest to oil palm – for various reasons the planting of oil palms was rather slow to begin with – but gradually, more or less the entire area has been converted either to oil palm or more recently also to rubber.

The next substantial dereservation in Okomu took place in 1992, at the time of the military regime of Ibrahim Babangida, when ORREL – taken over by the Michelin Group in 1981 – sought to expand its land. It initially planned to expand further west, on non-reserved community land. This, however, included Udo's own land and was now vigorously resisted by the residents of Udo; the *Oliha N'Udo* recalled that they employed both physical and spiritual means – i.e., charms and rituals – to stop Michelin's bulldozers.[53] Frustrated in its efforts here, the company turned its attention instead to Okomu Reserve and applied for a lease of 2,000 hectares inside the reserve, immediately adjacent to its old land. This land, however, had already been applied for by someone else: the Iyayi Group. This is the company of Efionayi Iyayi, a prominent businessman from Benin City and one of the largest business groups in Edo State, involved in timber logging, rubber plantations and other business enterprises. The Iyayi Group had arranged a large bribe to be paid to top government officials, but Michelin had even bigger financial muscle and also the backing of the *oba* of Benin. The *oba* had a business stake in nearly all companies, with operation bases in Edo forests, apart from Iyayi. Consequently, it was Michelin who won the lease for these 2,000 hectares, but the then administrator of Edo State, who had received considerable financial support from Iyayi in the past, promised him another 2,000 hectares of reserve land to compensate for his previous loss. Iyayi quickly had the area in question surveyed, but in 1993 when negotiations had been almost completed to issue Iyayi with allocation papers, the government was swept out of power, when Moshood Abiola won the elections. In the course of the subsequent annulment of Abiola's election, the setting up of an interim government and Sani Abacha's seizing of power in November 1993, various government reshuffles took place in Edo State, during which many of those opposed to Iyayi's acquisition of forest land were removed. This included the Head of Forestry at Iguobazuwa, the headquarters of Ovia South-West LGA, who was transferred to Benin Headquarters. Iyayi was therefore eventually granted his 2,000 hectares, just south of the land given to Michelin.[54] This was done officially in the form of timber allocations, and indeed Iyayi logged his compartments (and the neighbouring ones) extensively. However, de facto he acquired more permanent rights to his compartments and he started clearing and planting rubber in a number of them.

Later on Iyayi applied for and acquired a further 3,000 hectares of land south of his existing allocation and 1,000 hectares were given to J.O. Ighile, another Benin businessman.[55] In 2001 – Nigeria now under civilian rule and Lucky Igbinedion the Governor of Edo State – OOPC took over all of Iyayi's and Ighile's land, 6,000 hectares in total. Given that neither Iyayi nor Ighile had officially bought the land, this sale was in some ways quite illegal, but seems to have been conducted without the involvement of the state government or the Forestry Department. Similarly, OOPC expanded elsewhere by taking over land officially given to J.A. Ogbomo, a businessman and prominent local politician from Udo, with numerous political connections to Benin City. Through these semi-official purchases, Okomu expanded its territory to over 25,000 hectares – more than 20 per cent of the forest reserve.

At about the same time another businessman from Udo, J.I. Mojo, was granted four compartments (over 1,000 hectares) around Iguafole, just north of the national park, in order to establish a large-scale plantation of oil palms and food crops. In 2002, despite a ban on dereservation (Forestry Department 2002a), J.A. Ogbomo was allegedly allocated a further six compartments (1,560 hectares) around Igueze, a village on the northern border of Okomu Reserve. His land was being demarcated in January 2003 – before the forthcoming elections in April 2003, which could potentially jeopardise his acquisition of the land through changes in government. The town of Udo too was granted an additional 50 hectares of community land.

Similar processes have taken place throughout Edo State. By 2002 requests for dereservation approved since the 1970s amounted to 1,333,132 hectares, or 23 per cent of total reserve land (Forestry Department 2002b). This officially reduced reserve land to 21 per cent of Edo State by 2002

(Forestry Department 2002a). A small percentage of this land was given to expanding villages in enclaves or on the borders of reserves but the vast majority, over 95 per cent, was allocated to large- and medium-scale oil palm and rubber plantation projects (Forestry Department 2002b). These included not only OOPC and ORREL but also Presco, another Belgian-run oil palm plantation in Edo State. A new wave of land allocation took place in the final years of Igbinedion's rule. In 2006 large parts of Ehor Reserve were given to Iyayi, parts of Obaretin Reserve to Presco, and parts of Ogba Reserve were dereserved and given to communities.[56] In 2007, in the final months of Governor Igbinedion's rule, Michelin was allocated a further six square miles in Iguobazuwa Reserve, just north of Okomu, causing protest and outrage amongst local communities (Enogholase 2007).[57]

Many residents of Udo and villages in Okomu Reserve condemned the large-scale allocation of land.[58] Some stressed that all these allocations had taken place when the Ministry of Agriculture and Natural Resources (MANR) was under the charge of Commissioner Joseph Amowie, after the beginning of civilian rule and Igbinedion's two terms in 1999. The *Oliha N'Udo* argued that during the military era 'nothing like that' happened. This is not quite true: as we have seen, the allocations to Iyayi and Michelin were made during the military era. However, land allocation did increase during Igbinedion's rule, in particular in his last years in power.[59] It remains to be seen whether Adams Oshiomhole, the new Governor of the Action Congress (AC) who came to power in November 2008, will continue on this course or not. Whether during military or civilian regimes, it is clear that such allocation of land was largely possible, and indeed shaped by, patronage networks, with land allocated to political supporters and financiers. At Okomu, Iyayi gained his land deals through years of financial support to Benin governors, Mojo had built up ties to Benin political networks through loans, and Ogbomo himself held several positions of political influence.

Reserve land, however, has become the most easily available not only for large-scale projects; small-scale cash- and food-crop production too have expanded in Okomu Reserve and elsewhere. The most significant development here has been the expansion of farmland under the Taungya system, which will be considered separately in Chapter 5. But several other additional developments have occurred, including the growth of cocoa farming by Yoruba migrants. These started coming to Okomu Reserve from neighbouring Ondo State in the 1990s, when the government of Ondo State launched a campaign to evacuate illegal cocoa farmers from its forests. The majority of cocoa farmers settled at Mandoti in the western parts of Okomu Reserve, which borders Ondo State directly. By 2000 there were also various camps in the eastern parts of the reserve, with about 150 cocoa farmers each at Assamara and at 'new site', just north of OOPC, and various other camps of thirty or forty people or so at other places in the reserve.[60] There was also a camp of about twenty people in Iguowan enclave, who had been granted permission to stay there by the Iguowan community. Cocoa farms varied in size between five and twenty acres, depending on the numbers of labourers available; one labourer was needed for working one to two acres. Farms were usually organised in groups, in that there was one *balee* (headman) in charge of all the different farms of a camp, but each farm was then worked by different farmers and their labourers. They operated by starting to farm in one area and then expanding into another until the original camp was so far away that they had to build another camp there.

These Yoruba settlements were not legal, but the Forestry Department's efforts to eradicate this kind of illegal farming were hampered by staff shortages, logistic restraints and lack of support from the state government. Forest guards themselves, as people in Udo and villages would emphasise, were at times complicit in allowing cocoa farmers to stay, often without the knowledge of their superiors. In addition, the conservator in Iguobazuwa complained that even if they found and prosecuted cocoa farmers, the government did not move in to destroy their farms or buildings, so they usually just continued undisturbed.[61] In fact, Yoruba cocoa farmers had some protection from local Edo communities. Thus, the first cocoa farmers at Mandoti were invited to settle by the *onogie* of Obanyator,[62] who apparently in this way sought to check the advance of Ijaw farmers from the south.[63] In the same way that cocoa farmers at Mandoti were 'under the protection' of the *onogie* of Obanyator, those closer to Udo were under the protection of the *uwangue* of Udo, who had put a specially appointed *odionwere* in charge of them. Any new cocoa farmers who arrived had to be brought to see him, and he was responsible for 'sort[ing] out any quarrel and report[ing] it to the palace'.[64] Udo also provided some protection in their dealings with forest officers. In return for his services, he and Udo as a whole received some payment from the cocoa farmers. With such widespread support from local communities, a cocoa board set up in 2002 in emulation of the one in Ondo State, was quite ineffective. In fact, by 2006 it had rather given up on trying to prosecute cocoa farmers, and instead focused on collecting annual rent from them, part of which was supposed to be used for forest regeneration.[65]

Plantain farming, finally, spread in different ways again. It began to emerge in the late 1990s in the southern parts of the reserve around Mile Three. Mile Three and other settlements in the area are former logging camps, where former timber workers have stayed on to farm under the Taungya system. Under the Taungya system farmers normally plant one or two acres each with yam and other food crops for subsistence purposes. In the late 1990s, however, the Forestry Department became more 'lax' and was either willing to allocate more land, or increasingly unable to monitor farming activities in this area. Farmers took advantage of the greater availability of land and developed a new method of plantain farming. This new method was inspired by the success of Yoruba cocoa farmers at nearby Assamara. As well as producing cocoa, they derived additional income from selling plantain to the large and expanding Lagos market. Plantain farming requires much less labour than yam farming, as the forest does not have to be clear-felled and burned before the plantains are planted and less weeding is required. Consequently, each farmer can look after much larger areas of plantain, and by 2003 many farmers in Mile Three had farms of eight or more acres. On these they grew plantains, under a canopy of some high forest trees that were left standing. In the northern parts of the reserve too, several larger plantain plantations of this kind were established around 1999. As with the cocoa farmers nearby, these larger plantain plantations were divided between several owners, who each in turn employed a few labourers. These owners and their labourers came mostly from Eastern Nigeria – from Calabar, Cross River and Ibo states – but also from further afield, including at least one labourer from Cotonu.

The ways in which much of the forest of Okomu Reserve has been converted to other uses, therefore, include official government decision, in the case of the initial establishment of OOPC; under the table land allocations and sales, in the case of the ORREL and OOPC extensions and Mojo's and Ogbomo's land; illegal but locally endorsed and gradually accepted encroachment by Yoruba cocoa farmers; and the unofficial expansion and adaptation of the Taungya system by plantain farmers. It is right that the reserve has become a source of patronage: not just to the governor and other state officials, but also to local communities and forest guards receiving payments from cocoa and plantain farmers. These developments, however, in Okomu and elsewhere, have been possible precisely because reserves are under government and not community control. Just as control over reserves gave government officials the opportunity to allocate land to political cronies or to sell it profitably to paying expatriate plantations, so local communities welcomed Yoruba cocoa farmers and other strangers into the reserve more than they would have done onto their own community land, building up their own profitable patronage networks.

However, patronage is not the only logic behind forest conversion in reserves. Palm oil, rubber, cocoa and plantain production all present important sources of income as well as food and can form a vital part of rural development. Population growth, the country's worsening economy, stalling urban development and the liberalisation of exchange rates and ensuing rapid inflation as well as the reduction of agricultural subsidies under Structural Adjustment Programmes in the 1980s and 1990s all combined to create substantial increases in food prices. This not only made food-crop agriculture a profitable enterprise again for relatively well-off people; it also forced many to return to agriculture to survive (Okuneye 2002). In this context, Scott's (1969) argument about 'informal' policy making through corruption or law breaking is relevant, in particular to the ways in which small-scale plantain and cocoa farming have developed in Okomu Reserve. The reorientation of the Edo cocoa committee may be seen as an instance of policy making from below, with developments on the ground eventually becoming part of official policy.

The Social Effects of Land Conversion in Okomu Reserve

In the 1960s Udo was still small and more a village than a town. By the time of my research in the early 2000s, however, it was a bustling and rapidly expanding town of 15,000 or so inhabitants with a hospital, a library and several primary and secondary schools. Udo market was flourishing, attracting people from the whole region, the outskirts of the town were a constant building site with new houses and roads rapidly constructed, and new shops, such as a photocopying and a hifi shop, were opened every month. This development had been steered and encouraged by Udo chiefs, who, acutely conscious of Udo's history as once a major rival of Benin City, were keen for Udo to regain its former greatness. As Udo was expanding again, it did so along historical lines, according to the *Oliha N'Udo*, Jonathan Ezele. Thus, old town quarters that had long been abandoned came to life again one by one, so that in 2000 there were nine quarters, each headed by a different chief and each with its own shrines and allegiances. New streets were given the same names as streets in Benin City; Ezele, for example, whom I accompanied on several of his frequent tours through Oliha quarter in his battered old Mercedes, inspecting the development of new buildings and plots, had called two new streets in his quarter College Road and Uzama Street.

In their eagerness to make Udo grow again, Udo chiefs actively encouraged newcomers, or 'strangers' as they are generally called, to settle in Udo and the surrounding areas. Here, the *Oliha N'Udo* took a leading role. It has already been described, above, how he persuaded others to consent to the establishment of the Osse River plantation in 1964, and he subsequently persuaded them again to allow strangers to buy land and build their own houses in Udo. Other Udo chiefs were initially vehemently opposed to this, just as Edo settlements in general were reluctant to give up their community land to others. Oliha recalls a meeting during which he persuaded others of the advantages of inviting strangers in:

> People were afraid that strangers would take all their land, but I was the one who changed this. They were trying to make a law, around 1967, but I prevented it. I said, in a meeting: how can the strangers feed their children? Before, people thought that if the strangers had no land they would have to buy all the food and make them (the native) rich, but this is not so. Development comes from inviting strangers; they often bring development with them.

Consequently, strangers were allowed to settle and farm, but there are marked differences between the different quarters of Udo in this. Oliha quarter and others on the outskirts are full of strangers – the sale and rent of plots no doubt providing the *oliha* with a handsome income. Their land is heavily farmed, whilst older quarters, in particular Ogbe and Ogiefa quarter, have been more protective of their own land. Today they have better farmland and more fallow land at their disposal. In this way, Udo's different quarters mirror the different policies of other settlements.

Udo's growth has been greatly facilitated by its position on the border of Okomu Reserve and the processes of forest conversion that have taken place in recent years. For one, Udo has benefitted from Yoruba cocoa farmers and plantain farmers in the reserve, who, as described above, have been paying fees to Udo for being allowed to farm inside the reserve. Udo inhabitants have also been involved in and profited from cocoa and plantain trade, just like cocoa and plantain farmers themselves and other traders in the area. After the cocoa harvests – which began in the Udo area around 2000 – buyers from Udo and Iguobazuwa bought the cocoa and rented lorries to take it to Ibadan and Lagos. Due to a production crisis in Cote d'Ivoire, cocoa prices were rising in the early 2000s and cocoa farming was a profitable enterprise both for the Yoruba cocoa farmers and their buyers. Cocoa prices fell again after 2003, but rose sharply in 2008.[66] Plantain farming too was profitable. It is not as expensive as other forms of food production, and there has been much demand for plantains in urban centres like Lagos in recent years. In the early 2000s there was a thriving trade in plantains in the Okomu area, with several large lorries packed to the rim with plantains setting off daily from Okomu Reserve to Lagos. Another advantage of plantain farming in this respect is that plantains are less seasonal than other food crops and can be harvested almost all year round. Plantains presented an important source of income not only to their growers, who were mostly male, but also to their largely female traders from Udo and other places in the area. These market women bought plantains directly from the farmers, often walking many hours into the forest to collect the harvest. Organising the whole transport from inside the forest to Lagos, even small-scale traders could make ₦10,000 (£50) a week in 2001. The economic importance of the plantain trade for women can be ascertained from the fact that there was an association of all plantain traders, which any woman who wanted to trade in plantains was obliged to join.[67]

Another significant factor in Udo's growth was the establishment of OOPC and ORREL, which contributed to the local and regional economy by creating employment, building infrastructure and providing other benefits. In the early 2000s ORREL employed about 580 permanent staff, OOPC about 1,300. These had relatively good salaries and other benefits, including accommodation on the plantations. Many permanent workers came from other parts of southern Nigeria – Yoruba, Ibo, Delta areas and especially Cross River State. One reason for this, I was told, was that initially local people had little interest in working on the plantations; but by 2000, when OOPC had begun making profits, about 35 per cent were from Edo State.[68] However, the majority of workers were non-permanent staff, who without social security, housing or pensions were cheaper and could be hired seasonally. They were hired on a daily basis until 2003, when legislative changes made the employment of day labourers illegal. Since then, they have been hired as contract labourers. In 2006 about 80 per cent of labour at OOPC was secured through contracts.[69]

Many of the non-permanent workers lived in Udo, where they rented rooms and spent some of the money they earned. In general, Udo is the community that has benefitted most from the two plantations. In 2000 OOPC tarred the road going through Udo to the plantation, which greatly improved access between Benin City and Udo and its surrounding villages and attracted traders and businesspeople as well as farmers to the town. Contact between Udo and the plantation increased, enabling OOPC workers to visit Udo's market and shops more frequently than before. Udo's market and transport business flourished, and many new houses and shops were built. Both plantations have also been involved in community development projects: OOPC dug a few boreholes, provided maintenance for the town generator and made some donations to the healthcare centre and the library, whilst ORREL provided a generator and dug a borehole in Iguowan, the village most directly affected by its expansion.

In addition, both plantations, seeking to build up local support, were generous in their support of Udo leaders. OOPC gave the *Uwangue N'Udo*, the town's ruling chief since 1999,[70] generous monthly allocations of palm oil that he distributed amongst other Udo chiefs, whilst ORREL distributed firewood in the same way. The plantations also strategically awarded various outsourcing contracts. The *uwangue*, for example, had been put in charge of security at OOPC's newly acquired land in the east of the reserve, which gave him access to vehicles and other resources. Ostensibly their task was to prevent illegal logging, but in practice managers at OOPC were quite aware that this arrangement helped the *uwangue* to reap further profits from logging.[71] In 2000 OOPC also made a man from Udo Personnel Manager at the plantation, which meant he was able to secure jobs for some local people. Finally, OOPC vehicles, like timber and plantain vehicles, paid charges at Udo roadblocks, a system of taxation which, in one form or another, contributed to local development.

However, the very ways in which the plantations, and in particular OOPC, interacted with Udo fostered rising conflict in the town which, in 2006, led to the rejection of the *Uwangue N'Udo* as its town leader and the takeover of the town by 'the youth'. As I have explored in more detail elsewhere (von Hellermann 2010), the crisis was rooted in both ongoing chieftaincy rivalries and electoral political competition in the run up to the 2007 local election.

But OOPC contributed to this in several ways. When the plantation became more and more profitable, the *uwangue* began to try and exert an ever greater monopoly over contracts and ceased to redistribute palm oil and other benefits from the plantation. This caused resentment amongst Udo chiefs, potential contractors and others seeking work. The plantation's switch to contract labour further exacerbated these tensions, as it provided the *uwangue* with additional avenues to profit from OOPC and increased competition between potential contractors. Plantation management also gave exclusive support to the *uwangue*, whose increasingly oppressive and predatory rule of Udo was in part based on the knowledge of this support. Moreover, the plantation's tendency to pay large sums of money at roadblocks encouraged extortion practices. OOPC is not alone in this: multinational oil companies in the Niger Delta are also embroiled in local conflicts not just because of the high stakes they present, but also because of the particular ways in which they interact with local communities, often supporting the most predatory segments of society (Groves 2008).

The plantations not only contributed to Udo's political crisis, they also present a threat to farmers in Okomu Reserve. Some of the original cocoa farmers in the west of the reserve had made their farms on OOPC land that was still forested. When OOPC started expanding its planted area, these farmers had to make way and move elsewhere. In the same way, much of the new land OOPC bought in the southeast of the reserve in 2000 was occupied by cocoa farms. They were given some time to allow them collect their first harvest, but some parts have already been bulldozed by Okomu and the remaining cocoa farmers will eventually have to leave, too. The expansion of OOPC therefore had some direct detrimental impact on those farming – illegally but with local endorsement – inside Okomu Reserve.

Other local farming communities too have been negatively affected by the expansion of the larger plantations. The extension of ORREL in 1994 threatened to completely engulf the village of Iguowan, situated just on the northern border of the national park. It lost the land it was farming through the Taungya system at the time, and it was only through the campaigns of Iguowan villagers and an International Labour Organisation (ILO) representative who had come to Iguowan through the Nigerian Conservation Foundation (NCF), that the village remained at all. Likewise the land Mojo obtained is close to the village of Iguafole and includes substantial areas that were farmed by Iguafole residents under the Taungya system, whilst Ogbomo's land was farmed by the inhabitants of Igueze before he acquired it. Most recently it is farmers in Iguobazuwa Reserve – more than 20,000 people – who lost their land through Michelin's expansion, despite a forceful protest campaign that involved the World Rainforest Movement and Friends of the Earth.[72] The legal ambiguities under which small-scale farming has flourished and the absence of recognised tenure rights have therefore meant that small-scale farmers are vulnerable in the face of larger, more powerful interests who enjoy political support, even if these do not necessarily have clear legal rights in reserve land either. The gradual expulsion of small-scale farmers considerably reduces the opportunities that the accessibility of Okomu Reserve had provided to poorer, less well connected members of the rural population.

The Ecology of Forest Conversion in Okomu Reserve

Just as large- and small-scale farms have had different social outcomes, so their ecological effects too have varied. Whilst many aspects of the different impacts of forest conversion are still poorly understood (Donald 2004; Hartemink 2005), some assessments can be made on the basis of both my observations and discussions with farmers and of existing research into the relative impact of cocoa, plantain, oil palm and rubber production. These highlight not only that there are significant differences between different tree-crop plantations, but also that the actual environmental consequences of different processes of forest conversion are quite different from how they are often portrayed in Benin forestry circles.

The Forestry Department has described illegal cocoa farming as 'the quickest means of destruction of the natural forests' (Forestry Department 2002a), but this assessment reflects their political status as 'illegal' encroachers into the forest rather than their actual impact. Cocoa farms are comparatively small in size and do not change plant composition as radically as large-scale oil palm and rubber cultivation. Farmers only partially clear the forest and begin by planting plantain, orange and other trees to provide shade for the young cocoa seedlings planted in between (Fig. 5). Various food crops are also planted to begin with. Over the years cocoa dominates these farms more and more, but there are still numerous smaller forest trees and larger shade trees in between. Elsewhere too, it has been shown that for these reasons such small-scale cocoa agro-forestry has comparatively little negative impact on biodiversity – or on soil erosion, soil fertility and carbon sequestering – and that it can play an important role in biodiversity conservation (Franzen and Mulder 2007; Rice and Greenberg 2000; Schroth and Harvey 2007).

The environmental impact of plantain farms of the kind found in Okomu Reserve has, to my knowledge, not yet been researched. On plantain farms farmers leave an even larger number of high forest trees, which provide shade, moisture and 'manure'[73] for the plantains planted in between (Fig. 6). By softening rainfall, plantains themselves also help to prevent soil erosion. Moreover, according to farmers in Okomu Reserve, plantains store minerals and water and are therefore soil enriching. When they die and fall, all of this goes back into the soil, just when it needs it in the dry season. In this way, I was told, the plantain farms can keep producing for a long time.[74] Both cocoa and plantain farms, therefore, present forms of farming that do not completely remove and replace forest ecology; economic timber species too can be protected within them.

In contrast, the impact of the large-scale oil palm and rubber monocultures of ORREL and OOPC is significantly more drastic. During the preparation of a new oil palm or rubber plantation, the forest or any other vegetation is cleared completely, leading to temporary increases in soil erosion and loss of carbon storage, as well as a drastic decrease in overall biodiversity which prevents the future regeneration of economic timber species (Fig. 7). Large parts of Okomu Reserve are, by now, covered by fully grown oil palms and rubber trees (Figs. 8 and 9). Under full cover, soil erosion is again reduced and biomass is relatively high (however, in the case of oil palm in particular, carbon sequestering is not as high as claimed by advocates of the palm oil industry, as palm trees are eventually felled and destroyed).[75] If oil palms replace grassland, they may increase carbon fixation, but they substantially decrease it if replacing forest (Germer and Sauerborn 2008). Moreover, both oil palm and particularly rubber plantations – although less is known about these – have a substantial negative impact on biodiversity and soil fertility (Fitzherbert et al. 2008; Hartemink

Figure 5 A Yoruba cocoa farm in Okomu Reserve. Photograph taken by the author, 15 January 2002.

Figure 6 A plantain farm in Okomu Reserve. Photograph taken by the author, 14 January 2002.

2005; Turner and Foster 2009). ORREL and OOPC also use fertilisers and pesticides which pollute the soil.

It is interesting to briefly compare these two plantations to other forms of rubber and oil palm production in Edo State. Traditional palm oil production was very different, in that oil palms were not planted; instead, those that naturally sprung up on abandoned farmland were fostered and maintained for many years. This kind of low-intensity and small-scale palm oil production had far less negative environmental impact. By now there are, however, several larger oil palm plantations. A number of these have run into financial and other difficulties and have been abandoned, sometimes before oil palms were actually planted. For example, in Ohosu Reserve, situated northwest of Okomu Reserve and bordering Ondo State, 2,000 hectares were allocated for a demonstration plantation by the Nigerian Institute of Oil Palm Research (NIFOR). The land was cleared but only 100 hectares were planted before planting was abandoned.[76] Today it appears as a vast empty plain when driving through it on the Lagos–Benin motorway. Here, soil erosion and loss of biodiversity are particularly high, and biomass is now significantly lower than outside Ohosu Reserve. In comparison to oil palm plantations such as these, OOPC's impact is less drastic. OOPC and ORREL managers also argue that they contribute to conservation within Okomu Reserve by leaving some of their land under tree cover and by shielding Okomu National Park from incursion by small-scale farmers. There are, however, real limits to the conservation activities of the plantations, which will be explored in more detail in Chapter 6.

Figure 7 Newly cleared land within OOPC. Photograph taken by the author, 13 August 2006.

Separating Farm and Forest | 79

Figure 8 Oil palm at OOPC. Photograph taken by the author, 16 February 2002.

Figure 9 Rubber trees at ORREL. Photograph taken by the author, 17 January 2002.

Compared to small-scale, local rubber plantations outside Okomu Reserve, OOPC and ORREL do rather worse. A large number of the rubber plantations that dominated the landscape around Udo in the 1960s have since been converted to food-crop farms, due to falling rubber prices and population increases, but there are still quite a few old rubber plantations around. These have a large number of forest trees and other plants between them, and often appear almost like other mixed, young, secondary forest from afar. They also do not use pesticides or fertilisers, so overall have had far less negative environmental impact than ORREL and OOPC's rubber plantations. Where planted on previously open land, as many were, they have actually increased tree cover outside reserves. Interestingly, such old and seemingly uncared for plantations, with many different trees and shrubs in between them, can be highly productive, producing particularly well after years of rest (Bauer 1948). Since rubber prices started to rise again in 2003, tapping of these old farms has in fact been resumed.

Overall, the environmental outcomes of forest conversion in Okomu Reserve have not been completely destructive; even the large-scale monocultures contribute to biomass (and therefore carbon sequestering). But locally managed cocoa and plantain farms, just like rubber plantations outside the reserve, preserve substantially more of the existing forest and re-create it by fostering different trees in between tree crops, thus maintaining high levels of biodiversity. The widely condemned expansion of 'illegal' small-scale agriculture inside reserves has therefore been less destructive than it is conventionally portrayed. However, with large-scale monoculture plantations now rapidly replacing such local forms of agroforestry, their environmental as well as their social benefits are lost.

Conclusion

Forest reservation in the Benin Division was never the technical measure of rational forest management scientific forestry portrayed it to be – the simple protection of existing forest estate. Rather, it was a highly political process that brought about a fundamental reorganisation of both landscapes and land ownership that had far reaching consequences.

Benin landscapes had been characterised by symbiotic and fluid relations between farm and forest, but reservation divided farm and forest into separate spheres. In this way it protected areas of forests but not overall biomass or biodiversity. Moreover, it actually curtailed the regeneration of timber species, the very trees scientific foresters sought to protect and foster. This was because reservation forced farmers to abandon their previous methods of sporadically farming large areas and concentrate farming on smaller areas closer to villages, causing not only hardship to local communities but also terminating the key dynamic behind the regeneration of light-demanding timber species in Benin forests. At the same time, reservation fundamentally changed access to land by placing more than 60 per cent of land under government control. The fact that all reserves were put under Native Administration made little difference here; on the contrary, it gave the *oba* and the Benin Council far more direct control over reserve land than they had ever had, and meant that at independence, reserves came under the control of the government of the then Midwestern State.

In this way, reservation itself created the conditions for large-scale dereservation and forest conversion in recent decades. As this chapter has explored, demand for land for tree-crop production had been growing since the 1930s, all the more when late colonial and post-colonial government policies increasingly prioritised agricultural development. Because community hold on unreserved land has remained strong, despite attempts by the Nigerian government to break it (in particular through the Land Use Decree of 1978), it is reserves that have, in the end, become more easily available for large-scale plantation projects than community land. In the process, they have indeed become a source of patronage for politicians, just as those deploring the decline of forest management have argued. Large parts of Okomu Reserve have been given to political allies and donors in recent years, whilst on a smaller scale proximity to Okomu Reserve has enabled communities such as Udo to build up patron–client relations with illegal migrant Yoruba cocoa farmers, whom they have granted informal rights to farm inside Okomu Reserve. Meanwhile the breakdown of the Taungya system, through informal agreements between foresters and local farmers (which will be looked at in more detail in Chapter 5), has also enabled the expansion of plantain farms within the reserve.

However, just as reservation did not actually constitute successful forest management, so its recent weakening has neither been purely the result of management failures and patronage politics nor has it had the destructive effects usually attributed to it. Illegal cocoa and plantain farms in particular address real needs for land, and the arrangements between local communities, forest guards and migrant farmers present 'informal' policy making from below, in the absence of formal measures from above. Notably, cocoa farming has recently been formally recognised by the Edo State Forestry Department in response to these developments on the ground. Cocoa and plantain farms have contributed significantly to the livelihoods of migrant farmers as well as local communities, without completely destroying the forest: because they involve only partial forest clearance, biomass and biodiversity remain relatively high and timber species can be preserved.

Okomu's large-scale plantations have also addressed needs for palm oil and rubber and created jobs and income for local people and migrants from other parts of Nigeria. Environmentally, the replacement of forest with monocultures of rubber and oil palm also means that biomass has overall changed less than through forest removal alone. Biodiversity has been drastically reduced, however, and OOPC in particular has also contributed to recent communal conflict. Moreover, the plantations' expansion is threatening or has already brought about the destruction of many cocoa, plantain and food-crop farms, with little compensation for their owners. These developments highlight the vulnerability of small-scale farmers in the 'frontier' environment that Okomu has become, despite their comparatively positive role in local development and environmental change. Altogether, then, the history of reservation and dereservation in today's Edo State has been more intertwined and more complex than simple narratives of decline suggest.

Notes

1. Nigerian National Archives Ibadan (NAI), CSO 26 29697 V, Benin Forest Scheme, 437.
2. 'Osunbor begins probe of sale of government property by Igbinedion', *The Sun On-line*, 1 September 2007. Retrieved 23 February 2009 from http://sunnewsonline.com/webpages/opinion/editorial/2007/sept/01/editorial-01-09-2007-001.htm.
3. National Archives (NA), CO 879/69, Enclosure 3 in No. 138, Address by the Conservator of Forests, Southern Nigeria, before the Chamber of Commerce, Liverpool, 16 September 1904.
4. An interesting discussion of colonial preoccupation with customary land tenure, and its roots in British idealism, can be found in Cowen and Shenton (1994).
5. One reserve, the Gilli-Gilli Game Reserve, was in fact already created in 1907.
6. NA, CO 879/69, Enclosure 1 in No. 83, Report by the Conservator of Forests, 20 June 1903.
7. Nigerian National Archives Enugu (NAE), Legislative Council Papers 1912, No. 12, 'Annual Report of the Central Provinces for the Year Ended 31st December 1910, 18'.
8. This was by Order No. 397, under the 1908 Forestry Ordinance. The *Gazette* notification states that:

    > 1. All that area containing three hundred square miles (300 square miles) more or less situated in the Central Province of Southern Nigeria and bounded on the North by a line drawn from the junction of the Okwo River with the Siluko River to the junction of the Agwehen River with the Okomu River and thence to the town of Ikoru on the Ovia River, On the East by the Ovia River and the Owate Creek, On the West by the Siluko River, shall constitute a Forest Reserve within the meaning of the Ordinance and all the provisions of the Ordinance shall apply to such area.

    > 2. The game and fishing rights of the Native communities within the limits of whose occupation the Reserve is situated shall remain unextinguished and the rights of the present holders of Timber Licences covering the above area shall be respected.' NAE, Order No. 397, *Government Gazette* of April 24 1912, no. 27, 1048.

9. NAI, BEN PROF 4/3, 4/3/4, Intelligence Report of the Benin Division, compiled by H.N. Nevins, 1930.

10. One rubber plantation was established by the British timber firm Miller Brothers at Sapele in 1901 (Fenske 2010).
11. NAE, CSO 3/5/3, Sir Ralph Moor, HBM Commissioner and Consul General, to the Secretary of State for Foreign Affairs, 14 November 1898; CSO 4/1/1, Letter by the Acting High Commissioner Gallwey to Mr Chamberlain, containing a draft of the Forestry Preservation Proclamation 1900, 3 Aug 1900. See also Egboh 1985.
12. NAI BP 334/1, Conservator of Forest, Benin Circle (BC) to Senior Conservator of Forest, Olokomeji, 3 May 1915; BP 628/1915, Agricultural Report for Half Year ending 30 June 1915, 1.
13. NAI BP 490/18, Resident (BP) to Secretary (SP), 27 November 1918; BP 725 Forestry offences prosecution for: Procedure.
14. Gilli-Gilli was already created as a game reserve in 1907, but existed more nominally than in practice. It was made a formal forest reserve in 1927.
15. NAI BP 103/1925, Letter from Oba Eweka to the District Officer, 5 April 1929; BP 80/29, Letter from the Director of Forests to the Chief Secretary of Government in Lagos, 2 November 1929. See also von Hellermann and Usualele (2009).
16. NAE, CAL PROF 53/1/558, J.N. Oliphant: A Further Report on Forestry Development in Nigeria, 1934.
17. NAI, CSO 26/30802, 5, Comments on the points raised in the Tabular Statement attached to the Honourable Chief Secretary's Letter to No. 30802/3 of 21 July 1936, 5.
18. NAE, CAL PROF 53/1/558.
19. Ibid., 5.
20. NAE, CAL PROF 53/1/558, 5.
21. NAE, Sessional Papers of the Legislative Council No. 1 1938, Annual Report of the Agricultural Department 1936, 26.
22. NAI, BP 41, Vol. V, Annual Report of the Benin Division, 1936.
23. NAI, BP 1132A, Petition by Benin Citizens Against the Proposed Forest Reserves, Yesufu Eke and 74 others, 13 September 1935.
24. NAI, BP 999, Letter from Oba's Council to the District Officer, 10 June 1935.
25. NAI BP 1273, Vol. II, Report on the Rates of Wages and Conditions in African-Owned Rubber Plantations in Benin Division by J.G.C. Allen, 1944, 355.
26. NAI, BD 207/124, Petition by Odighi Villagers, regarding the farming interests of Edionwe at Odighi, 17 April 1941.
27. NAI, BP 999, 41, Letter from Oba's Council to the District Officer, 10 June 1935; BP 999, 43, Letter from Oba to the Resident of Benin City, 28^t August 1935.
28. NAI, BP 1223, 2, SACF (BNA) to Resident (BP), 20 April 1936.
29 See also NAI, BD 207/73, Conservator of Forests, Benin Native Administration (BNA), to the Resident, Benin City, 6 January 1939.
30. NAI, WP 6056, Vol. II, WLG (W) 9, Okomu Forest Reserve.
31. Ibid.
32. Ibid.
33. NAI, WP 6056, Vol. II, WLG (W) 9. Urezen is still today famous for its Ovia shrines.
34. Interview with Chief Sunday Omoregie, the *obasoyen* of Udo, Iguowan, 14 February 2002.
35. All reserves were divided into compartments of a square mile each, which were used for logging allocations. At Okomu the numbers of compartments are still widely used as references and for orientation.
36. Interview with Chief Ekhator Uwadia, the *esogban* of Udo, Udo, 29 November 2002.

84 | *Things Fall Apart?*

37. Address by the Conservator of Forests, Southern Nigeria, before the Liverpool Chamber of Commerce, 16 September 1904, National Archives (Kew), Colonial Office Records, CO 879/69, Enclosure 3 in No. 138.
38. *Daily Times*, 15 May 1936 and *Nigerian Eastern Mail*, 'The Benin Forest Reserves', 25 June 1938.
39. NAI BP 1470/Vol. II, Permanent Crops in Benin Division: Planting of Order No. 28 (No.1&2) 1939; ibid., 119, Ag. Resident (BP) to Secretary (WP), 2 April 1941.
40. NAI, BD 1381, Vol. II, letter from the Catholic Mission to the District Officer of Benin, 27 July 1943; unidentified newspaper article.
41. NAI, BEN DIST 4, BD 672/416, District Officer to Oba of Benin, 24 June 1943.
42. NAI, BEN PROF 1, BD 27/Vol. X, Annual Report of the Benin Native Administration Forest Circle 1943.
43. The villagers of Odighi, for instance, petitioned against an 'enterprising individual', Chief Akenu, who stemmed from Odighi but lived in Benin City. He had taken hold of large tracts of land to supply Benin City with yams and cereal crops, ousting villagers from it. NAI, BEN DIST 6, BD 207/124, Petition by Odighi Villagers, Regarding Farming Interests of Edionwe at Odighi, 17 April 1941.
44. See also NAI, BD 24, Vol. I, Benin Native Administration Forestry Protection.
45. NAI, BD 27/Vol. X.
46. NAI BP 999, 132, Joseph Olotu and E.O. Amayo, representing Uyere Court group of villages and people of Odighi, Odiguetue, Ugboki, Onwan and Agbelikaka to Resident (BP), 19 May 1944; NAI BP 990, 49, G.O. Amayo and S.I. Idemudia of Odighi Village to Western House of Assembly, Ibadan, 28 April 1953.
47. NAI, BD 27, Annual Report on the Benin Division 1945, 5.
48. NAI, BP 1470 Vol. II, 162, H. Spottiswoode, Ag. Resident, Benin Province to Secretary, Western Provinces, 26 June 1948.
49. These principles are based on a letter by Herbert Howard, Inspector General of Forests, India, printed in the *Empire Forest Journal*, Vol. 23, No. 1, 1944.
50. Annual Reports of the Forest Department, 1952–1962.
51. After independence the official name of the department changed from Forest to Forestry Department.
52. The other three were located at Ayip-Eku in Cross River State, Ore-Irele in Ondo State and Ihechioma in Imo State (Egbon 1990).
53. Interview with Chief Oliha, Udo, 20 January 2003.
54. Source unknown. Cited in the 1995 Master Plan for Okomu Forest Reserve.
55. Interview with an OOPC Manager, August 2006. See also 'The Edo Property Scam', *The Source Magazine On-line*. Retrieved 17 September 2007 from http://www.thesourceng.com/edopropertyscamsept17.htm.
56. Interviews with forest officers at Benin City, August 2006. For an overview of land sales under Igbinedion, see 'The Edo Property Scam', *The Source Magazine On-line*.
57. See also World Rainforest Movement, 'Tyres at the expense of people's livelihoods'. Retrieved 17 February 2009 from http://www.wrm.org.uy/bulletin/138/viewpoint.html#Nigeria.
58. Interview with Chief Oliha and several other Udo people, 15 August 2006.
59. 'The Edo Property Scam', *The Source Magazine On-line*.
60. These are rough estimates given to me by the *odionwere* of the cocoa farmers in January 2003.
61. Interview with the Conservator of Forests, Iguobazuwa, 8 January 2003.

62. The *enogie* of Obanyator was one of the titles created when parts of Iyek'Ovia, previously entirely under Udo, were carved out and put under the control of other title holders. See Eweka (1992).
63. This was general knowledge in the area. It is also mentioned in Adeleke (1999) There are long-standing antagonisms between the Ijaw and the Bini in the area.
64. Interview with the *odionwere* of cocoa farmers, Udo, 3 January 2003.
65. Interview with members of the Cocoa Committee, MANR, Benin City, 29 August 2006.
66. See http://www.icco.org/statistics/monthly.aspx, viewed 4 March 2009.
67. During my fieldwork in 2002 a successful tradeswoman in Udo tried to circumvent this association and independently bought a load of plantains, but this was discovered and she was taken to the police station. When she returned to her lorry, which she had already loaded, all the plantain had become rotten, so she lost around N150,000 (£750).
68. Interview with the Personnel Manager at OOPC, 21 December 2002.
69. Interview with an OOPC manager, 14 August 2006.
70. Traditionally, as discussed in Chapter 2, Udo has been ruled by the *Iyase N'Udo* since the defeat of Udo by Benin in the early sixteenth century. The last *Iyase N'Udo*, however, came into conflict with the *oba* of Benin in 1999 when he sought membership of the Council of Traditional Rulers, which of all Benin chiefs only the *oba* was a member of. Since this constituted a direct challenge to the *oba*'s power, he was ostracised by the *oba* for seven years and stripped of all his powers. In his stead his younger brother, the *Uwangue N'Udo*, was appointed as ruler of Udo. For a fuller discussion of Udo politics, see von Hellermann (2010).
71. Interview with an OOPC manager, 10 October 2002.
72. See http://www.wrm.org.uy/countries/Support_to_Nigerian_Communities.html, viewed 24 February 2009; see also http://ecologicalequity.wordpress.com/themes/stories-of-right-stories-of-might/434-2/ and http://www.foei.org/en/resources/publications/annual-report/2007/what-we-achieved-in-2007/member-group-victories/africa/building-up-community-forest-management, both viewed 11 July 2012.
73. 'Manure' is the word generally used by people when speaking to me in English.
74. Interview with plantain farmers in Okomu Reserve, near Mile Three, 20 February 2002.
75. See Rhett Butler, 'Indonesian palm oil industry tries disinformation campaign', 8 November 2007. Retrieved 25 February 2009 from http://news.mongabay.com/2007/1108-palm_oil.html.
76. See, for example, http://ucheanudu.tripod.com/id3.html, viewed 25 February 2009.

CHAPTER 3

Managing the Forests
Logging and Regeneration

Introduction

Just as large parts of forest reserves have been converted to other uses today, so forest management itself – the regulation of logging activities and tree regeneration – has virtually collapsed in recent decades. Working plans have been abandoned, plantations are not maintained, and logging levels – legal and illegal – are high. Most timber merchants, loggers and forest officers agree that Edo forests are now virtually 'finished' and that timber exploitation can only continue for a few more years. In many ways this indeed presents a Nigerian 'things fall apart' story: for decades Edo State was the subject of the most intense schemes of scientific forest management in Nigeria, and economic crisis, corruption and lack of political interest have played significant roles in its recent collapse. Yet again this story is more complicated than simply an abrupt departure from a once well-functioning system of forest management. For one, as with forest conversion, there are a range of different arrangements between forest officers and loggers today, and not everyone is complicit in corruption in the same way; there is even some limited resistance to it. More importantly, recent developments again need to be understood in continuation from previous phases of forest management, with the roots of many of today's problems not in its recent but its longer history.

This chapter therefore again delves deeply into history. It explores forest management from its beginnings in the early twentieth century, when logging concessions first began to be allocated; throughout the 1920s and 1930s, when scientific experiments with tree regeneration were set up and the first ambitious working plans drawn up; through large-scale application of working plans and natural regeneration under the Tropical Shelterwood System (TSS) in the 1940s and 50s; through industrial forestry and artificial plantations in the 1970s and through its decline since the 1980s. It shows that throughout this long history, scientific forest management in fact never achieved its central goal: the creation of a regular, large and sustainable timber supply. This goal consisted of two parts: the development and control of Nigeria's timber industry, and the fostering of sustained regeneration of economic timber species. The balancing of these two potentially quite contradictory aims lay at the heart of scientific forestry, but was, in practice, difficult to achieve. The Nigerian Forest Department never gained the necessary control over either forests or logging activities: its regeneration attempts only had limited success, and for years it struggled to expand timber demand, only to see it eventually taking off quite beyond the department's control. Moreover, the timber industry became an integral part of Benin's existing patrimonial political economy, with title holders and other 'big men' soon recognising timber as a lucrative source of income.

This chapter then looks at recent developments in the logging industry and its regulation. It explores how, in continuation from practices that had developed over the course of the century, the allocation of logging concessions has recently been determined largely by political networks and personal connections, just like the allocation of reserve land for plantation projects. But the logging industry today also involves many smaller operators with less political clout. Altogether, a range of different 'legal' and 'illegal' arrangements now exist between loggers and forest officials, which are explored in detail. Not all forest officers, however, participate in these dealings, whilst some local communities have developed strategies to regulate logging and plant trees themselves. In addition to bringing out these important nuances, recent developments are altogether put in perspective by relating them directly to the legacies of previous forms of forest management. Viewed as the last stage of a series of different, often unsuccessful approaches rather than an abrupt break from proper management, forest regulation's recent decline emerges not as a consequence of recent management failures but as virtually inevitable.

'Supervision and Control': 1900s–1910s

In 1903 H.N. Thompson was appointed as the first Conservator of Forests in the Protectorate of Southern Nigeria with the mandate to set up a system of scientific forest management. He was chosen for this task on the basis of his twelve years of experience in the Indian Forest Service in Burma, where he had been 'able to pass through and become more familiar with all the necessary stages in forest organisation from its introduction to its completion'.[1] In 1903, however, Thompson was still some way away from introducing scientific forestry proper to Southern Nigeria. He had to build up the department's staff, gather information on the extent and nature of the forests he was to manage, and, as we saw in the last chapter, gain government control over the forest estate. For the first two decades, therefore, forest management itself was limited to the 'supervision and control'[2] of concessionaire logging activities and local forest use.

Timber logging in West Africa began in the 1880s, when European and American markets that had previously relied on mahoganies from South America discovered the qualities of West African mahogany (*Khaya ivorensis* and, to a lesser extent, *Khaya senegalensis*).[3] A strong hardwood of attractive red colour, it was sought after for furniture making and especially for interior panelling on ships and railways, production of which was rapidly expanding at the time. A temporary falloff in mahogany supplies from British Honduras in the early 1890s provided a further opportunity for West African timbers to become established in European markets (Dumett 2001). Several European logging companies, as well as some African firms, began operating in Cote d'Ivoire, the Gold Coast and the Lagos Colony at this time (Callahan 1985; Dumett 2001; Moloney 1887). Amongst these, the forests of the yet inaccessible Benin Kingdom acquired something of a mythical status and were rumoured to be rich in mahogany as well as rubber (Anene 1966: 144). Following its conquest in 1897 several European companies immediately applied to the High Commissioner of the Protectorate of Southern Nigeria for permits to begin logging in the Benin area. These included a Mr Bleasby, the African Association, and Messrs Alexander Miller Brothers (Anene 1966: 297). Miller Brothers were one of the oldest trading firms in West Africa and had already been operating in the Gold Coast for some time (Dumett 2001).

High Commissioner Moor and other colonial administrators had ambivalent feelings about the concessionaires' eagerness to exploit Benin forests. On the one hand, as a strong believer in the benefits of free trade (Afigbo 1970), Moor was ready to exploit and develop the timber resources of the Benin forests, which he saw as one of the main tasks of the Forest Department.[4] The administration needed European firms as the main agents of the desired 'opening up' of the Benin Kingdom and as a source of revenue. On the other hand, Moor and his colleagues were rather wary of concessionaires, whom they perceived to be rapacious and profit driven – an attitude which they shared with many colonial administrators of their generation (Heussler 1963; Nicolson 1969; Thompson 2005). Thus, upon receiving the requests for logging concessions in the Benin area, the Acting Commissioner wrote to the Foreign Office: 'What they would probably do would be to work their concessions with great energy so as to get as much out of it [sic] in the shortest possible time and having drained these resources, to seek concessions elsewhere … the Natives of the Protectorate … would be left with their country very much the poorer in certain products' (Anene 1966: 289).

Having seen the destruction in Lagos Colony caused by unregulated logging as well as rubber tapping, Moor was highly concerned that Benin forests might be similarly destroyed if concessionaires were not regulated and controlled by the government. He therefore established a Forest Office and appointed a first provisional Forest Officer, P. Hitchens, as early as 1899. In the same year, he and Hitchens drew up the first provisional forest regulations, which set out a regulatory framework that, formalised in the 1901 Forest Proclamation, largely remained in place for several decades. Overall it was based on a licensing system, whereby 'all persons wishing to engage in lumber work should be required to take out a licence, the cost of which should be fixed ... that they may be under control as to the work and their competency as workers'.[5] Licensing simultaneously generated income for the Forest Department and enabled it to supervise and control logging activities. In the course of applying for a license, concessionaires were required not only to give evidence of their financial resources, but also to give detailed information on each individual tree they were intending to cut (Unwin 1920). After examination of stumps by forest staff, concessionaires had to pay export duties on each tree exported.

The regulations also stipulated that concessionaires should pay royalties to the 'owners of the soil'.[6] Royalties were divided between paramount chiefs and the nearest villages to a concession, who each received half (Igbafe 1979). This ensured that local people too profited from logging, a measure that was by no means taken in all colonies where concessionaire logging was encouraged (Bennett 2000: 321). Nevertheless, Moor's approach here is in marked contrast to his agricultural policy, where he actively encouraged local small-scale production and prohibited European companies to set up large plantations.

The location, size and length of concessions were also regulated. The 'Rules Relating to Timber' of the 1901 Forest Proclamation stated that concessions were to be granted for seven years at a time and were to measure nine square miles, with one mile of water frontage each (Egboh 1985). There were also stipulations about girth sizes of the trees to be felled: 'the dimensions of timber that may be felled should be determined in diameter and circumference in accordance with the growth of the various timbers'. In this way 'the younger growing trees will not be cut before attaining in or about the maximum growth'.[7] Finally, 'each licence holder should be required to plant out in suitable positions a certain number of seedlings in accordance with the number of trees felled'.[8]

These measures were designed to balance conservation concerns with commercial interests. It was not easy to find a workable system that would achieve this, however, and the regulations were continuously revised under a rapid succession of new forest ordinances in 1903, 1905, 1908 and 1916. For example, under the 1901 ordinance only trees of girth sizes of nine feet or over were allowed to be cut.[9] In his first forest surveys in 1903, however, Thompson noticed with alarm that the age distribution of trees in the forest was very uneven: most trees were 'over-mature', and there were few intermediate trees. Consequently, he proposed to raise the girth limit from the previous nine to twelve feet, to ensure that enough mature trees were left to produce seeds (Thompson 1906). The logging companies Thompson Blois and Kjellgren & Company protested that there was no market in Europe for such gigantic trees, but the colonial government insisted on a twelve foot girth limit for mahogany, cedar and walnut (the bulk of the trees cut at the time) compensating for this increase by also increasing the size of concessions to 100 square miles. However, by 1913 it was reduced to ten feet again, in order to encourage more felling (Dennett 1913). By the time of the Forestry Ordinance of 1916 different categories of trees were established, with different girth size stipulations for each (Unwin 1920). In all this, it must be noted, girth size regulations were based on rough estimates, rather than on exact scientific calculations.

There was also continuous struggle between government and concessionaires over appropriate fees, royalties and export duties, as the Forest Department was trying to maximise revenue without making license costs prohibitively high (Egboh 1985). Describing these negotiations in some detail, Egboh notes that whilst government revenue increased considerably, royalties payable to chiefs remained constant at ten shillings per tree cut (ibid.: 97). Fee regulations were also becoming increasingly complicated, with different fees payable for each class of trees (Unwin 1920). Likewise, the procedures by which licences were obtained, girth sizes controlled and payment of fees and royalties ensured, were also repeatedly discussed between the government and the concessionaires. The latter frequently protested at the cumbersome and obtrusive inspection procedures and the administrative work required of them (Egboh 1985: 96).

Nevertheless, the establishment of the concession system facilitated a quick rise in the issue of logging licences. After the passing of the 1903 rules, logging applications were so numerous that by the end of 1904 practically the whole of the Benin territory had been applied for twice over (ibid.: 101). The main companies, at this time, were Alexander Miller Brothers, Messrs Bey & Zimmer, and Scott McNeill & Co, amongst others, the biggest being the Miller Brothers of Glasgow. Many of these, including the Miller Brothers, operated along the Siluko and the Osse River and in the area in between that became Okomu Reserve in 1912. In 1907, when almost all the land of the Benin territories had been leased to European firms, the *Civil Service Handbook* stated confidently that 'the fullest development of this important industry is assured'.[10] Between 1907 and 1912 on average around 5,000 trees were cut each year in the Benin Division, accounting for over 80 per cent of all trees felled in the protectorate. Concessionaires also largely planted, as obliged, the twenty tree seedlings for every tree felled; in 1907, 63,565 seedlings were reported to have been planted, and 71,768 in 1912 (McLeod 1908; Thompson 1913).

However, whilst logging was administered and regulated by the Forest Department, in the sense that most participants more or less followed the rules of applying for licenses, paying fees, facilitating stump inspection and planting tree seedlings, this did not mean that the department actually had real control over logging activities and that thereby 'the fullest development of this important industry' was indeed assured. In reality, the extent of the department's control over the industry was rather limited. Overall, the location, nature and extent of logging activities was not determined by the department but rather by a combination of topographical, technical and economic factors beyond the Forest Department's control. The main reason why the Benin District was by far the most important timber producer of Southern Nigeria throughout this period was its extensive river system, which made access and log transport much easier than in the Western or Eastern Province (Adeyoju 1966). At this time trees were cut manually by axe and then rolled on tree stumps towards the nearest river, from where logs were floated towards Lagos or other markets, each operation involving a considerable number of men (Fig. 10). Logging activities were therefore restricted to within three to four miles of the rivers, as high labour costs for hauling out trees did not make extraction economically viable beyond this point. The introduction of tramways from the 1910s onwards, much welcomed by the department, slowly increased the range of logging activities, but these were always ultimately determined by cost.[11]

Figure 10 Early logging in Benin forests. Taken from Richard St. Barbe Baker, *Africa Drums*, 1942, London: Lindsay Drummand Ltd.

The market, as a whole, proved to be more difficult to stimulate and expand than the department had hoped. Every year new species were sent to Kew for testing and identification and to Liverpool merchants in the hope of finding buyers. These attempts, however, had only limited success. By 1913, Dennett wrote, 'Iroko has at last established itself in the European market under the name of "African Teak"' (Dennett 1914). But mahogany remained by far the most important species, which limited logging activities considerably. According to the Forest Department's annual report of 1907, 'many new firms, anxious to get a footing into the timber business in Southern Nigeria, applied for tracts of country which they had never examined, and only after they had obtained provisional sanction to work, did they discover that there was little or nothing of present value on these areas' (McLeod 1908). In that year twenty-one new timber areas were granted, 'but most of these were untouched either because some were granted at the close of the year or because they were found to contain little or no mahogany' (ibid.: 245).[12]

Moreover, the mahogany market itself was quite volatile. It was found in 1909 that '[o]wing to the poor prices obtainable for mahogany during the early part of the year, the leaseholders of timber areas restricted their operations' (Thompson 1910). In 1910 an increase of tree felling permits – from 2,889 in the previous year to 4,684 – was attributed to 'better prices prevailing in the home markets' (Thompson 1911a). Then the First World War greatly upset the Nigerian logging industry. In 1916, due to the prohibition of the import of timber into the United Kingdom, only 825 trees were cut for export (Thompson 1917). The Admiralty's demand for mahogany, in particular for the construction of aeroplanes, helped for a little while to give an impetus to 'what was a declining industry', and the number of areas exploited rose again in 1918 (Thompson 1919). But when the Admiralty contracts closed in 1919, tree export declined once more and of the 11,200 square miles leased out, only about a third was actually worked, much to the Conservator's dissatisfaction (King-Church 1920). Overall, in Nigeria as throughout the tropical world, the lack of demand for tropical hardwoods meant that timber production remained relatively low. In Cote d'Ivoire and Central America too the focus on mahogany alone meant that only a few trees scattered through the forests were cut, and loggers searching for them were referred to as 'mahogany hunters' (Gonggryp 1948). In fact, up to the 1930s tropical hardwoods made up less than 10 per cent of the 50 million cube metres of wood traded in world markets, and their volume was so insignificant that it was not recorded in international trade statistics compiled by the Comite International du Bois (Bee 1991). In Nigeria the value of timber exports was minimal compared to that of palm kernels and palm oil; between 1909 and 1912 timber constituted only 1.35 per cent of all exports, whereas palm kernels and palm oil's share constituted almost 93 per cent.[13]

Similarly, the department was not as successful as it may have wished at fostering local development. Logging operations did, of course, provide income-earning opportunities for the many workers needed for the extraction of timber, but overall numbers were still quite low and the wages of log workers are never mentioned in contemporary records as a contribution to local development. The contribution of timber royalties, in contrast, is discussed; the 1907 Annual Report, for example, states that 'the Chiefs and villages interested in the various timber areas obtained £1,169.5s 0d. each in respect of royalties, and are so well satisfied with the manner in which the timber rules are worked, and the large sums of money handed over to them, that it is reported that they assist in inspecting and counting stumps, in order that no trees may be felled that have not been paid for' (McLeod 1908). However, the image of villagers all enthusiastically participating and gaining from the logging industry is a little misleading. Royalties were not only small compared to government revenue, they were also unevenly distributed. Even if villages received half of all royalties, it is likely that these would have benefitted male elders and village chiefs rather than the village as a whole. The main beneficiaries were the paramount chiefs, who in addition to royalties also received rent from the European firms established at Siluko and Benin, the profits from the sale of rubber plantations in the villages, and official tribute and rents paid by non-natives of Benin (Igbafe 1979: 266). Altogether, as Igbafe puts it, 'the paramount chiefs had tremendous opportunities for power, influence and wealth' (ibid.: 134). For example, Agho Obaseki, the first *iyase* of Benin City under colonial rule, had several villages surrounded by particularly rich timber-producing forests under his control, some of the first to be exploited. Consequently, Obaseki made more money from timber royalties than any other chief (ibid.: 123). His son, Gaius too, who later succeeded him as *iyase*, became a rich timber producer (Bradbury 1973b). It was also exclusively chiefs who were beginning to become timber traders themselves at this time, an opportunity all others were excluded from.[14]

The Forest Department's attempts to regulate local forest use were exploited by chiefs in a similar way. Colonial regulations of local use were informed by the same two aims as those of concessionaire logging, namely conservation and revenue creation. Foresters believed that without control and regulation of every activity, local forest use practices would lead to irreversible forest destruction. At the same time, they saw all forest products as natural resources and forest use as economic activities that, through regulation, could be harnessed towards the overall project of revenue generation and local development. As with concessionaire logging, these two goals could both be happily addressed through a permit and licensing system: it would both regulate forest use and generate revenue. To begin with the extent of such regulations was limited, but this changed with the Forestry Ordinance of 1916, which stipulated that licences had to be acquired and fees paid for a large number of trees and minor forest produce. Since people depended on the forest for a wide range of foods, medicines, tools and building materials, this profoundly affected every aspect of local people's lives, and, unsurprisingly, evoked a large amount of protest:

> Petitions were received complaining of the hardships caused by the number of trees protected against damage from farmers and by the tax on minor forest produce other than rubber, more especially that on palm wine, bamboo poles, fibres, dyes and, in the case of timber, on iroko trees cut for the manufacture of wooden bowls, pestles and mortars; with regard to these complaints it may be remarked that as a large trade is done in all these forest products the plants from which they are derived are as much in need of protection as the rubber yielding species. No doubt a great deal of the agitation was of the kind one would naturally expect when any untaxed article is taxed for the first time, and no great importance need be attached to it. (Thompson 1917: 11)

To some extent, the department responded to these protests by changing some regulations:

> [T]he grievances were inquired into and wherever there were reasonable grounds for complaint administrative steps were taken to remove them pending an amendment of the laws. It was found expedient to limit the number of trees protected against damage from farming to those in the first class of the schedule,[15] to apply the Regulations relating to fuel only to cases in which it is used on a large scale such as for driving machinery or for brick burning; to exempt petty traders from its operations and to reduce the number of protected minor forest products'. (Ibid.)

Despite these adjustments regulations continued to apply to a large range of forest products, and with increasingly rigorous enforcement the number of prosecutions grew considerably (Thompson 1919: 5). Virtually criminalising all normal farming and building activities, the regulations were highly irksome for local people, and there were repeated protests against them. Nevertheless, they remained in place for many years, long after the target of 25 per cent of land under forest reservation, when they were supposed to have been lifted, was reached. Perhaps one reason for this was that the numerous permits required for felling trees resulted in 'a satisfactory increase of revenue' (Thompson 1919: 5). They also presented an additional source of income for the paramount chiefs, who were in charge of enforcing these regulations. They employed agents whose job it was to collect permit and license fees. These were highly unpopular either because they were trying to enforce the new laws, or because they abused their position for personal gain (Igbafe 1979: 169). Most local resentment focused on agents, but it was the paramount chiefs, and the government, who benefitted most from the licensing system.

The Forest Department also initiated the setting up of communal mahogany plantations in villages, often near rubber plantations (Dennett

1914: 7). Like rubber plantations, these plantations would have been under the charge of paramount chiefs, who were responsible for the recruitment of labour. As on rubber plantations, this labour may well have been enforced, rather than voluntary. By 1920 about forty-seven Benin villages had such plantations (Unwin 1920: 167). At the same time, the government started its own mahogany plantations (Thompson 1917: 14), as well as several teak (*Tectonia grandis*) plantations in the Ishan plateau.[16]

The ecological impact of concessionaire logging, regeneration efforts and restrictions on local forest use was probably quite limited at this time. Certainly, a considerable number of large mahogany trees were removed or 'creamed off', but their numbers and geographical distribution were constrained by the fluctuations of the market and high extraction and transport costs. Forest regulations of local forest use, even if vastly irksome, would also not have considerably changed overall patterns of use. Tree planting too was rather ineffectual. Most companies carried out their duties in only the most perfunctory way, so a large number of seedlings died quite quickly (Egboh 1985: 99; Dennett 1914). Generally, concessionaire planting was not regarded with optimism by the Forest Department. Communal and government plantations also proved disappointing, as they all suffered from severe pest attacks and were soon abandoned. On the whole, it is hard to tell how many of the tree seedlings planted in this early phase survived, though perhaps quite a few more than thought at the time.[17] There may indeed have been a certain professional prejudice in the department's dismissive attitude towards concessionaire planting, for in the 1920s Thompson noted with some surprise that, in several places, these seedlings had become tall mahogany trees (Thompson 1926). However, all these efforts were so small scale and localised that they probably did not have a significant regenerative impact.

Far more significant were the particular ways by which the logging industry and forest regulations became lodged into the local political economy. For one, in contrast to agriculture, logging itself was organised almost entirely around foreign concessionaire interests, with few local logging companies participating. There were some gains for local people through wages and royalty payments, but royalty payments were few and unevenly distributed, as is so often the case in such systems (Larson and Ribot 2007). Whilst paramount chiefs became rich through timber royalties, normal villagers did not receive anything. Perhaps most importantly, it was in these years that the overall organisation of the Benin logging industry, namely the concessionaire and licensing system, was set up. This system, whilst widely used throughout the logging world, has also always, like many other systems of regulation through licensing, been prone to abuse and corruption at all levels: throughout the history of scientific forestry there have been forest guards open to bribes, and government officials allocating concessions in return for extra payments or favours (Rochel 2005). In the Benin area this system has remained in place largely unchanged to this day, and is central to understanding the logging industry throughout the twentieth century – in particular, perhaps, to understanding developments in most recent decades, when in a sense forestry largely returned to the basics of its beginnings. In the intervening decades, however, it was mostly the trials and tribulations of scientific forest management itself that preoccupied foresters.

From Sapoba Research Station to the Benin Forest Scheme: 1920s–1930s

From the 1920s onwards the British Colonial Service was becoming increasingly professionalised across a range of sectors, as 'the diverse assortment of military officers, adventure seekers and occasional psychopaths who mingled with the early generation of proconsuls was chased out in favour of an earnest cohort of professional functionaries schooled for their service'(Young 1988: 49). Forestry was foremost amongst these, partly perhaps because it was one of the 'first loves' of Ralph Furse, the Colonial Service's influential recruitment officer (Heussler 1963: 147), but also because, in the wake of acute timber shortages experienced during the First World War, the building up of reliable timber supplies from the colonies had become a priority for the British government (Worboys 1979). It was now that empire forestry properly developed. An annual Empire Forestry Conference was set up in 1920, the journal *Empire Forestry* was launched in 1921, and in 1923 the Imperial Forestry Institute was founded at Oxford.[18] The empire forestry conferences were attended by representatives from throughout the empire, and became the main platform for exchanging experiences and developing empire-wide policies (Powell 2007; Rajan 2006).

For the Nigerian Forest Department these conferences gave important direction and policy initiatives. H.N. Thompson attended the 1923 conference in Canada, at which urgent calls were made for the introduction of proper scientific forest management in all colonies. Returning from the conference he wrote with some impatience in his Annual Report that '[s]ylviculturists are urgently needed for the purpose of investigating the best methods of regenerating the Forests and until the required data have been collected little progress can be made in exploiting our more valuable Forests under working plans suitable to the local conditions of the country' (Thompson 1925: 8). Indeed, as a new generation of foresters arrived, some were rather frustrated by the lack of actual forest work being done at the time. In his memoirs St. Barbe Baker, newly posted from Kenya in 1925, recalls writing to a friend: 'I'm issuing permits for a vast amount of timber to be felled and have practically nothing to spend on reforestation or the practice of scientific forestry, which I regard as my real job' (St. Barbe Baker 1942: 90).

According to St. Barbe Baker – very much a self-publicist – his frank words triggered an avalanche of telegrams that eventually led to the decision to establish a proper research station. Given that Thompson himself was keen to set up a research station, it seems unlikely that St. Barbe Baker played such a central role, but at any rate, in 1926 a research station was indeed established in Sapoba Reserve in the south of the Benin Division, with initially St. Barbe Baker in charge (St. Barbe Baker 1928: 203). Following Thompson's plans, St. Barbe Baker set about demarcating a 'forestry working area' of 600 acres that was divided into four blocks, each containing five departments of 30 acres. He started trials with five methods of forest treatment, under each of which different plants were cut back or suppressed by other means in order to facilitate the growth of economic species (ibid.: 204–5) (Fig. 11). His enthusiasm for his work – often, he claims, keeping him up until midnight, reading the latest forestry journals and silvicultural reports from India – is reflected in the recollection of Sapoba in his memoirs:

> Back in the woods on a large clearing are hundreds of highly scientific experiments in progress to determine the best means of perpetuating for all time the wealth of the woods. Man has interfered, hence he must continue to assist Nature to restore what she has yielded to man. Minute records are kept of what Nature is doing, in hundreds of experimental quadrates, and each month a team of trained foresters observe, count and chart tree immigration following silvicultural operations. In the past the equilibrium of the forest was upset; for years, exploitation by itself was the vogue. Then came a great awakening. This last best mahogany forest was threatened, and in time would have gone the way of others. The process has been reversed. Forest workers have been trained to carry out the most advanced cultural operations known in the forestry world. The tide of destruction has been turned and the well-being of the Kingdom of the Wood has been restored. (St. Barbe Baker 1942: 101)

These words not only reflect St. Barbe Baker's own enthusiasm, but also convey the hopes and expectations of scientific forestry at the time. However, his trials were not to play the important role St. Barbe Baker envisioned. In 1927 he was replaced by J.D. Kennedy, one of two fully trained silvicultural specialists who were now finally posted to Nigeria (Ainslie 1929: 8). Kennedy continued with the experiments begun by his predecessor and regularly supplied detailed reports on their progress to the Forest Department Headquarters in Ibadan. But, always limited in scope by lack of funds, the acute financial shortages the Forest Department suffered between 1930 and 1933 as a result of the Depression brought about a virtual standstill in silvicultural experiments at Sapoba, and the methods experimented with here were soon forgotten (Mutch 1952: 12).

98 | *Things Fall Apart?*

Figure 11 Silvicultural experiments at Sapoba Research Station. Taken from Richard St. Barbe Baker, *Africa Drums*, 1942, London: Lindsay Drummand Ltd.

Whilst Sapoba Research Station therefore did not immediately live up to the expectations of its founders, international pressure to build up proper forest management systems in tropical forest areas remained strong. In 1929 Chief Conservator Ainslie had reported from the First International Silvicultural Conference at Stockholm that 'all forests which are under exploitation should be placed under some form of working plan', and that the Congress 'would wish Governments responsible for tropical and subtropical forests to study the lines of advance taken by British India' (Ainslie

1930: 7). Major Oliphant's two visits to southern Nigeria, in 1933 and 1934, were part of this renewed drive towards increasing and securing global timber production through scientific forest management. As described in the last chapter, his visits resulted in the setting up of the Benin Forest Scheme, a pilot scheme to establish scientific forestry in southern Nigeria. The Benin Native Administration Forest Circle[19] was chosen for this project not just because of the *oba*'s willingness to cooperate, but also because it was already the most heavily exploited area in Nigeria.[20]

The Forest Scheme's primary aim was to set up proper working plans – as Thompson had already outlined in 1904, these were the central strategy of scientific forest management. If lack of personnel as well as forest reserves had prevented Thompson from ever setting up and implementing working plans himself, by now, with large areas of forest reserved and a growing cadre of foresters, working plans could finally be tackled. A sense of what these consisted of can be gained from the *Memorandum on Working Plans*, produced for the Forestry Conference held in Lagos in 1937. It describes the enormous amount of work necessary in order to secure a 'constant supply of younger plants in all stages of growth and in necessary proportion ...year by year for all the time':

> [T]he framing of even a simple working plan demands an intimate knowledge of the requirements of all species of plants (not trees only) dealt with, their rate of growth, their methods and times of propagation and the best treatment to secure such propagation. This requires survey and enumeration work in the field. It involves laboratory research and usually years of work on special areas, after which preliminary working plans are drawn up and applied for a series of years but subject to periodic revisions to allow for necessary modifications due to altered conditions or advance in knowledge.[21]

Okomu Reserve was one of the first in which this rather ambitious project was embarked upon. On the basis of an enumeration survey started in 1933 and completed by 1935, the Benin Circle[22] Conservator R.A. Sykes and the Senior Assistant Conservator of Forests W.D. MacGregor decided that the Nikrowa forest in the southeast of Okomu Reserve would be the most suitable in which to make the first working plan, as this district had 'some of the richest forest in the division and probably in Nigeria'.[23] In addition, a permanent sample plot was established at Nikrowa in August 1934, measuring seventy acres in total, on which different silvicultural methods were tried in a similar fashion to those at Sapoba. The positions of all trees in the plot were carefully mapped and measured, and the important species, a total of 1,413, were selected for remeasurement, with the object of preparing volume and increment tables. The plot was creeper-cut (i.e., creepers and other undergrowth were removed) and divided into 7–10 acre strips, all treated slightly differently.[24] This sample plot was to form the basis from which to develop a working plan. In addition, R. Ross, who had done most of the enumeration work in this reserve, also wrote a description of Okomu Reserve, 'with an account of the factors, historical, ecological and economic which will have to be taken into consideration when the plan is prepared',[25] and began drafting a preliminary working plan for the reserve. The overall idea was that once working plans for individual reserves were in operation, these would be joined together in larger units. For this purpose, in 1938 the Benin Native Administration Forest Circle was divided into three divisions: the Central Division (HQ Benin), the Osse River Division (Nikrowa) and the Jamieson Division (Sapoba). Each of these was to be managed under a working plan, becoming a complete forest administrative unit (Weir 1938, 1939). However, the realisation of these plans was soon hampered by problems.

One key difficulty was to secure the cooperation of logging companies. Whereas in the sample plots at Sapoba and Nikrowa pre-exploitation treatment was undertaken by the Forest Department, it was clear that for large-scale working plans to be operational in the long term, they would need the cooperation of the logging companies, who would be responsible for cutting non-timber as well as timber trees as and when prescribed by the working plans.[26] However, logging companies were reluctant to agree to this. Expatriate logging companies significantly decreased in number at this time, as many had come into financial difficulties after the 1929 financial crash. Consequently, almost all firms had merged or been bought up by the United Africa Company (UAC), which now held fifty-one concessions of ninety-three in southern Nigeria (Ainslie 1930: 19).

There were also a few local contractors, the largest of which was Gaius Obaseki, the *iyase* of Benin. When Akenzua II succeeded his father to the throne, in 1933, he campaigned vigorously to have more concessions allocated to local contractors. He was no doubt partly motivated by the fact that, by strategically allocating concessions amongst Benin chiefs, he could use concessions as a way of building up his political power. Timber concessions also presented an important source of income for the *oba* himself. Nevertheless, he was right that the proportion of allocations going to local contractors was minimal compared to that going to large expatriate firms. Whilst colonial foresters recognised that there was 'very little forest at the disposal of independent native exploiters',[27] they were highly reluctant to increase local concessions. Purportedly, this was because of the 'unskilled and wasteful cutting',[28] the 'waste and ineffiency'[29] of local logging practices, and because local contractors' 'non-payment of labour has given more trouble than anything else during the past few years' (Weir 1939: 12). The persistent condemnation of local practices,[30] however, may well have stemmed from the fact that, for the Forest Department, it was easier to work with one large logging company when trying to coordinate logging allocations and working plans; certainly, they were far less concerned with the prevention of monopolies than before. Whilst through

Akenzua's persistent campaign the number of local allocations increased, it still remained small.

But even dealings with favoured large expatriate companies such as the UAC were complicated, due to ongoing difficulties with both the overseas and the local market. As Thompson had already noted in 1922, 'a satisfactory system is difficult to introduce unless communications between the reserves and the more important local markets are established and a greater demand for secondary timbers arises' (Thompson 1923: 9). The department tried its best to create a local market, but local demand for timber remained disappointingly low throughout the 1930s (Weir 1939: 13). It also made great efforts to expand the range of timbers attractive to the European market. For this purpose, specialist officers in 'forest utilisation and wood seasoning' had already been appointed in 1928. After a survey of the existing sawmill operations, they tested a range of woods for durability, termite resistance and other qualities, and identified several as useful substitutes for the softwoods being imported for use in government departments at the time. Obeche (*Triplochiton scleroxylon*) seemed particularly useful, and, growing abundantly in many parts of southern Nigeria (Ainslie 1930: 23), soon became the most important export species (Adeyoju 1966). These efforts to find and promote the use of new species of wood were revived under the Benin Forest Scheme; a timber testing laboratory was erected in 1938, and in the same year Nigerian woods were taken to the Glasgow Empire Exhibition, where the Nigerian government supplied the inlaid floors and panelling for the royal apartments used by the king and queen on the opening day (Weir 1939). The introduction of new sawmills, such as the UAC sawmill in Sapele, was also expected to 'be the means of introducing new species to the home market as it is believed that small parcels of sawn timber will be bought as an experiment where the same purchaser, with a conservative outlook, might not risk the purchase of logs' (ibid.: 12).

Despite these efforts, the tastes of the European market remained 'fastidious'; in 1939 Oliphant reported that still only five timbers 'may be regarded as established market favourites, namely obeche (*Triplochiton scleroxylon*), African mahogany (*Khaya ivorensis*), sapele wood (*Entandrophragma cylindricum*), African walnut (*Lovoa klaineana*) and iroko (*Milicia excelsa*)', which made up 92 per cent of all exports (Oliphant 1940: 14).[31] In addition, the market fluctuated considerably; in 1929 the world market slump reduced the number of trees cut in southern Nigeria for export from 8,459 in the previous year to 2,420, as firms slowed down their work to wait for an improved market (Ainslie 1929, 1930). In 1938 it was again badly affected by political unrest in Europe, so much so that many timber camps were closed down.[32] Timber exports picked up with the outbreak of war in 1939, but demand remained confined to the same small number of species.

The continuing low levels of local demand, the 'fastidiousness' of overseas tastes and the fluctuations in global demand meant that even though some timber firms, such as the UAC, British West Africa Timber and the Nigerian Hardwood Co., were prepared to cooperate with the department to a certain degree, they refused to commit to detailed, long-term extraction plans.[33]

An even greater immediate problem, however, was the shortage of staff caused by the outbreak of war in 1939. After all the work of preparing working plans first for Okomu Reserve and then the Osse River Division, the Assistant Conservator who was supposed to be stationed at Nikrowa to implement it never materialised.[34] By 1942 there was only one European officer in the whole of the Benin Division, so 'forestry work was ... confined to administration and revenue-collecting, research and regeneration work being reduced to a minimum.'[35] For all these reasons, already in 1940 J.N. Oliphant, by then Conservator of Forests in Nigeria, reached the conclusion that the working plans drawn up so far were not appropriate. It is worth quoting this in full:

> It is now realised that past attempts at framing working plans of an elaborate type were premature and misdirected. They assumed a stability in both utilization practice and silvicultural method that is far from having been attained in Nigeria, and powers of control over timber cutting that would conflict with the economics of the traditional form of exploitation, and to some extent with the rights of concessionaires. As in other parts of the Colonial Empire where forestry is in a transitional stage, it has been found that a plan made at considerable labour and expense may within a year or two become out of date and useless owing to factors not foreseen at the time of its compilation.
>
> Forest use in a country such as India is highly developed, varied and complex, and needs planned regulation of an elaborate kind, whereas in West Africa it is still in its infancy. Planning is necessary, of course, but the need is rather for simple management plans covering large areas and providing primarily for the protection of the forest estate than for complex working plans designed to control working of a more intensive character than this region has yet seen. The basis for such simple planning is best provided by vegetation surveys carried out in more or less detail according to local requirements, and such work would have been of much greater practical value than the large-scale enumerations on which so much money has been spent. There is no need to dwell on past mistakes of this kind, but they are worth recording because colonial forestry has suffered so much from attempts to apply Indian precedents in circumstances to which they are not appropriate. (Oliphant 1941: 2)

In fact, these early, highly elaborate working plans were quietly abandoned over the course of the Second World War. When the war was over, a new, quite different attempt was made.

Technocracy, TSS and the Timber Boom: 1940s and 1950s

In the late 1940s a new age began in Nigerian forestry. Timber demand, so long disappointingly low, finally took off when a large amount of timber was needed in the rebuilding of post-war Europe, and local demand too began to grow. Key factors in this were the introductions of the chainsaw and of timber lorries, which together significantly reduced production and transport costs and made timber affordable for a broader range of consumers within Nigeria and abroad. At the same time, forest management itself took new forms, reflecting broader changes in colonial thinking. The earlier principle of self-reliance of the colonies gave way to an emphasis on colonial development, coming to fruition with the Colonial Development and Welfare Act of 1940 (Cain and Hopkins 1993; Phillips 1989; Young 1988). Like other sectors elsewhere, the Nigerian Forest Department had significantly more resources available than before, which made a considerable expansion of training, research and technocratic planning possible. There was a particular emphasis on the training of African staff, for the overall aim was now to work towards the eventual 'handover' of the forest administration, once Nigeria achieved independence. Moreover, as mentioned in the previous chapter, there was now a clear priority given to agricultural over forestry needs, which meant that scientific forestry itself changed to a more intensive form on less land. These aims together shaped the Ten Year Forest Administration Plan that the then Chief Conservator of Forests F.S. Collier drew up in 1945. The plan outlines a ten year programme for moving towards the final objective, 'the creation of a Forest Estate, under simple plans for a sustained yield of forest produce, staffed by an African cadre capable of development to assume entire responsibility for its management and adequate for a Nigeria fully occupied by a greatly increased population, with an improved standard of living' (Collier 1948: 8).

Over the next few years the plan was diligently put into practice. The Ibadan Forestry School, founded in 1941, ran an additional course for forest assistants from 1948, and that same year one Nigerian entered Nottingham University to read a degree in Botany and two Nigerians of the Junior Service, E.I.O. Akpata and E.M.O. Chukwogo, enrolled at the Imperial Forest Institute in Oxford. Upon their return, Akpata and Chukwogo were the first Nigerians to enter the Senior Service as Assistant Conservators of Forests; by 1951, 21 per cent of the Senior Service staff was Nigerian. At the same time, science and research became more organised. A Silvicultural Assistant was appointed in 1949, with the task of organising silvicultural records and of coordinating and recording all new research, and in 1951 a special Forest Research Branch was established at the old Forest Headquarters at Ibadan (Collier 1950, 1951; Rosevear 1952a). In 1952 Chief Conservator Rosevear also brought to life a new *Information*

Bulletin, in order to 'disseminate and build up knowledge of true forestry in the Department'(Rosevear 1952b: 2–3). In general, the language, presentation and organisation of Nigerian forestry became more technical and professional. Thus from 1943 Annual Reports included 'Standard Forms' containing a wide range of detailed statistical information. Because of the vast amount of paper and typist's time required, the standard layout 'was not enthusiastically welcomed by the Department', but Collier insisted that 'it must be persevered with and improved by use' (Collier 1950: 10).

Collier's plan also reorganised forest management itself. Nigeria's administrative provinces were priority ranked 'in accordance with the importance and vulnerability of their forest cover, with the requirements of the people and with the progress in Forestry already made in them'; the Benin Division was ranked first (Collier 1948: 8). For each province new working plans were drawn up which now covered the whole province, rather than single reserves. Like Thompson and St. Barbe Baker before him, Collier postulated that 'the object of any plan must be to convert the Benin forests from the "unmanaged" state with a very long rotation, to a "managed" state with a reasonably short and economically profitable rotation, to secure the Benin Native Administration maximum revenue from export, whilst supplying all local needs, in perpetuity' (Mutch 1952: 22–24).

His overall approach to drafting working plans was considerably simpler than the complex plans of the 1930s. For a start, it rested on the large-scale application of only one regeneration system: the Tropical Shelterwood System (TSS). In 1943 the forester G.M. Somerville was posted to Nigeria from Malaysia when this country was invaded by the Japanese. Somerville had many years of experience of rainforest regeneration techniques and was stationed at Sapoba Research Station, where he started off a new experimental series by poisoning trees around new obeche seedlings (ibid.: 19). The Nigerian version of TSS he eventually developed consisted of five rounds of climber-cutting or cleaning, two poisonings and two regeneration counts pre-exploitation, and three rounds of cleaning and two poisonings post-exploitation, followed by more regeneration counts. This system must have had positive results at Sapoba, for it was soon applied to most reserves in the Benin Division and beyond.

At the same time, for the overall coordination of logging activities Collier designed a simple compartment system with a rotation of 100 years, so that no more than 1 per cent of the total area of reserved forest would be felled annually. Thus, the Benin Division was divided into four periodic blocks of 25 years, during each of which 375 square miles, a quarter of the total estimated 1,500 square miles' forest estate of the Benin Division, would be logged – 15 square miles annually. Under the first working plans for Periodic Block 1 (1946–1969), timber companies such as the UAC (or the African Timber & Plywood Company, as it was now called) were obliged to notify the department in advance of their logging plans, to rest large parts of their concessions in order to give the department time to carry out silvicultural operations, and to cooperate closely in regeneration programmes in the future. For example, they agreed to a 'merchantability clause', which required 'the exploiting firm to pay for all timber defined as merchantable of some fourteen major species listed in the agreement, whether extracted from the forest, left at the stump or left as standing trees' (Collier 1951: 11).

Whilst it would seem that these stipulations restricted logging companies more than ever, this was not quite the case. Collier in fact solved the old problem of securing the cooperation of timber companies by persuading them to concentrate for the next few years on 'salvage felling', the uncontrolled exploitation of timber in unreserved land. This not only allowed the Forest Department to carry out necessary pre-exploitation treatment within the reserves, it also presented an easy loop hole for overcoming the problems existing legal rights of the concessionaires might have caused in changing the system within the reserves. Moreover, it was argued that salvage felling would enable timber companies and Native Administrations to 'salvage' all the economic trees on unreserved land that would otherwise simply be destroyed in farming operations (Collier 1951: 23). With the cooperation of logging companies finally secured in this way, the area under working plans rapidly expanded: by 1951, 76 per cent of the total reserved high forest of the Western Region were under working plans relating exploitation to regeneration – 'an extremely creditable performance'(Rosevear 1953: 3).

The consolidation of forest management coincided with an unprecedented timber boom. Between 1948 and 1949 timber, lumber, veneer and charcoal exports – almost entirely originating from the Western Region, and the majority of timber from Benin Province – were valued for the first time at more than a million pounds (Collier 1950: 4). By 1950, exports of logs and lumber had reached 'an unthought of volume': over nine and a half million cubic feet, almost double the previous year's record total, were exported and continued to rise to 'unbelievable' levels over the next few years (Rosevear 1952a: 2; 1953: 4, 34–5). For the first time in the history of the Forest Department, with a newly imposed export duty on timber, total revenue exceeded expenditure in 1951 (Rosevear 1953: 1). Whilst the timber boom slowed down a little in the middle of the 1950s, an important threshold had been crossed in that there were now no more reasons for foresters to complain about lack of demand. Both international and domestic markets were strong, with the domestic market in particular growing rapidly from this period onwards (Adeyoju 1966). The timber industry as a whole developed rapidly. The most significant change came through the introduction of the chainsaw in the early 1950s, which considerably reduced the amount of labour and time previously needed for cutting trees.[36] Transport too was improved through the introduction of timber lorries, such as the Austrian-made Steyr, still widely used today. Together, these developments made it more economical to increase the range of species and quality of wood extracted and greatly expanded the geographical span of logging operations (ibid.: 103). At the same time, the number of sawmills in the Benin Division rose from one in 1944 to seventeen in 1964, and that of sawmills in southern Nigeria as a whole from seven in 1944 to eighty-two in 1964. A large number of these processed wood deriving from the Benin Division (ibid.).

Accompanying these success stories, however, was a continuous undercurrent of worries and problems. For a start, the timber boom itself brought challenges. Whilst stealing and bribery had always been a feature of the timber industry, they now became more common as everyone's eyes opened to the potential 'gold mine'[37] the Benin forests presented. Logs were poached from rivers and the boom attracted a great number of speculators.[38] As Chief Conservator Rosevear wrote at the time, 'the trouble is that there is big money to be made at the moment in timber and the amount which unscrupulous dealers can afford to pay out for underhand deeds is large enough to be a very powerful temptation' (Rosevear 1952a: 15). Illegal loggers also frequently entered the compartments allocated to the UAC. Thus, whilst the timber boom was certainly highly welcome to the Forest Department, it was largely brought about by factors external to the department itself, and the department only had limited control over it. The more timber was recognised as a valuable resource and gained importance in the local economy, the less the department could regulate extraction.

At the same time, working plans and forest regeneration encountered problems. The large-scale forest administration plans covering the whole of Nigeria proved, in many cases, too ambitious and were soon abandoned, in particular in the north and east of the country (Collier 1951; Rosevear 1952a). Elsewhere too, large-scale application of TSS was difficult to maintain. TSS' complexity – involving thirteen different pre-and post-exploitation operations – caused numerous logistical problems, in particular with regard to procuring the necessary labour. Whilst there was already competition for labour between silvicultural works and nearby timber camps,[39] in the 1950s more and more people left rural areas for the 'congenial life of towns', making it increasingly difficult for the department to hold forest labourers (Rosevear 1953: 4).[40] By 1953, concerned that labour shortages would cause the complete breakdown of the system, forest officers reduced TSS operations from thirteen to six or seven (Robson 1955), whilst ongoing financial worries about the costs of TSS led to further simplifications of the system in the late 1950s. Even with an increased budget and higher revenues the Forest Department was constantly strapped for money and resources, in particular in comparison to the large timber firms operating in the area at the time. Donald

McNeil, director of the UAC in Nigeria from the late 1940s to the early 1960s, recalled the activities of the Forest Department as essentially rather ephemeral. He acknowledged that there was cooperation between the UAC and the Forest Department and that working plans were followed. But, he claimed, the bulk of the enumeration and other work was done by the companies themselves, who had far better technical and labour resources.[41] This is also emphasised in the UAC's 1958 film, *The Twilight Forest*, celebrating its timber operations in Nigeria, which displays a confidence in technological advancement, science and planning surpassing even that of scientific foresters.

Even at the height of scientific forestry, therefore, when the Forest Department had finally implemented proper working plans, conducted large-scale TSS operations and accrued some revenue, it still had not achieved the level of control over either logging or regeneration activities it was aiming for. But more fundamentally, it was also not achieving its overall ambition of matching high levels of yield with equally high levels of regeneration – the central aim of scientific forestry generally and TSS specifically. Whilst there were some signs of successful regeneration following TSS treatment, especially in the Benin Province (Henry 1956: 3, personal communication 2007), this was not the case everywhere. In Ijebu forests in Ondo Province, for example, no new seedlings of *Meliceae* came up at all (Rosevear 1952a: 20–21). The underlying problem, as Chief Conservator Rosevear saw it, was how 'intrinsically poor these so-called rich forests are with less than one useful tree to the acre and a yield that is never more than a twentieth and often as low as a fiftieth of what would be expected from a European forest … [E]ven the "rich" forests of Benin have less than a tenth of the volume of good timber to the acre that they should have' (ibid.: 6–7). Subsequent studies conducted by Richard Lowe and others in the 1960s and 1970s concluded that 'there was practically no management advantage in applying TSS treatments' (see also Lowe 1994; Okali and Ola-Adams 1987: 293; Parren 2003). Some later studies revealed that, in the longer term, TSS may have had more positive effects on the regeneration of economic species and also on biodiversity than previously assumed (Kio 1978; Okali and Ola-Adams 1987). However, as Okali and Ola-Adams acknowledge, it is not easy to discern to what extent regeneration in treated forests is the result of treatment or undirected natural regeneration. Often, untreated logged forest regenerates well naturally. In southern Nigeria J.C. Mallam reported in 1953 of a regeneration count of a 'sacrificial block' in the Gambari Group Reserve, in which no TSS operation had been conducted, and writes that 'quite unexpectedly, ample regeneration of the economic species has established itself within the dense tangle during the past four years' (Mallam 1953). Recent studies in the Budongo Forest Reserve in Uganda also concluded that regeneration of mahoganies after logging operations largely depended on pre-logging forest composition and not on arboricide treatment, which had little effect (Plumptre 1996). Overall, therefore, TSS no doubt contributed to forest regeneration, but it did not fully transform forests to the extent they wished.

What about the effects of logging itself? In general, they depend on the composition of the forest logged, logging intensity (rotation cycles, number of species and size of trees cut) and the temporal and spatial scale according to which effects are measured (Berry et al. 2008; Haworth 1999; Steege 2003). The impact of selective logging is now thought to be perhaps less destructive than previously assumed, with both economic timber regeneration and general biodiversity the same as or at times even higher than in unlogged forests (Berry et al. 2008; Cannon, Peart and Leighton 1998; Verburg and van Eijk-Bos 2003). Given that Benin forests were already the product of a long history of disturbance, colonial logging, even at increased volume in the 1950s, may not have significantly reduced biodiversity and possibly in itself contributed as much to the regeneration of economic timber species as TSS. However, even with some limited regeneration, virtually all of the largest standing lagos wood, iroko, ochebe and other trees were removed from the forest. Chief Conservator Rosevear recognised but also defended this in this telling passage:

> The fault of the forests is their lengthy history of disuse. There is no orderliness in them and they are rich only in the sense that an uncropped fruit tree in its season is for the moment rich. The mature and often over mature trees of today can be of little or no use to coming ages. They will have rotted or gone. It is they that occupy too much of the forest and the sooner they can be got rid of the sooner can the forest be more fully stocked and brought to a logical regularity. (Rosevear 1952c)

In fact, almost half a century earlier, H.N. Thompson had already made similar observations about the 'very unsatisfactory' stock of trees in southern Nigeria, typical of unexploited tropical forest – whilst there were large old trees and plenty of small seedlings, the intermediate gradation was 'practically absent'.[42] Both Thompson and Rosevear clearly recognised that economic trees do not regenerate well in undisturbed high forest. However, since neither TSS nor any other attempts by the colonial Forest Department ever successfully stocked up the forest or brought it to a 'logical regularity', the sum result of colonial logging in reserves was that forests were increasingly emptied of most of the economic timber species that had grown there as a result of previous forest and farm dynamics.

Moreover, a significant part of logging actually took place outside reserves. This was especially the case in the 1940s and 1950s, when salvage felling constituted 80 per cent of all trees cut (Rosevear 1952c: 5). Chief Conservator Rosevear noted that it was the prevailing high prices that made long and difficult journeys to tree stumps that had previously been uneconomic viable, but nevertheless wrote with some amazement: 'noone would have supposed twenty years ago that Nigeria's present vast export of timber could be furnished, as it is, very largely by "salvage" trees. There are doubtless great numbers of useful and valuable trees scattered throughout the forest belt though never in sufficient concentration or covering a sufficient area to make reservation possible' (Rosevear 1952a: 25–26). Both Collier and Rosevear stressed repeatedly that exclusive reliance on uncontrolled salvage felling by the timber industry during these years was a necessary measure, providing the time and space needed for silvicultural pre-exploitation treatment inside the reserves whilst rescuing valuable timber from destruction through farmland expansion: 'The present timber boom is certainly making use of a large amount of "salvage" timber but whether it is having any appreciable effect upon shortening the period for which such timber would last is open to some doubt' (ibid.).

This defensive remark conveys a certain uneasiness about the effects of salvage felling. Indeed, by 1952, trees outside reserves had started to become exhausted (Rosevear 1954). Contrary to Rosevear's vague assertion that salvage felling might have no 'appreciable effect', it drastically reduced the large amount of timber trees scattered in farm and fallow land. Such radical felling in unreserved land also disrupted seed dispersal and reduced opportunities for the regeneration of these economic species in bush-fallow farmland, already significantly curtailed by the growing intensification of farming on non-reserved land. Overall, therefore, despite the Forest Department's best attempts, forest management at this time was far from sustainable, with tree cutting far outpacing regeneration either in or outside reserves.

Industrial Forestry: From Natural to Artificial Regeneration, 1960s–1970s

If for decades domestic timber consumption had been considerably lower than the colonial Forest Department liked, from the mid 1960s onwards the opposite was the case. Now, as Nigeria's population grew and became increasingly urban and prosperous, the securing of an adequate timber supply to meet demand for timber for construction and furniture making was a pressing government concern. This was all the more the case after the discovery of oil, which brought with it the expectation of rapid growth and industrial development. Moreover, as part of the drive towards industrialisation, domestic paper production in newly built pulp mills was started, for which an additional large supply of wood was needed. Under these circumstances the Forest Department increasingly turned against TSS, which had already been perceived as inefficient for some years and had been gradually scaled down.[43] By the middle of the 1960s TSS, having 'failed to produce the results hoped for', was quietly abandoned (Lamb

1966); instead a 'more obvious' form of forest regeneration was needed (Lowe 1994: 38). It may well have been that larger political considerations rather than actual failures of TSS shaped these decisions; the development and modernisation ideals of Nigeria's newly independent government also influenced forestry, and forestry globally turned towards artificial regeneration at this time (Amanor 2001). For all these reasons, Nigerian forestry now adopted artificial regeneration through direct plantations. The agro-forestry method of Taungya planting too, which will be looked at in more detail in the next chapter, became more widespread at this time.

Artificial regeneration through plantations had been experimented with and conducted on a small scale in southern Nigeria throughout the twentieth century. These plantations included the mahogany plantations of the early twentieth century discussed above, a fuelwood plantation near Ibadan and various experimental teak (*Tectonia grandis*) and gmelina (*Gmelina arborea*) plantations at Sapoba and Olokemeji research stations. They had, however, been costly to maintain and also riddled with numerous diseases, so had so far been relatively insignificant as a reforestation method. In the 1970s, however, high oil revenues made a it possible to finance the large expansion of forestry staff necessary for the establishment and maintenance of large-scale plantations. The switch from natural to artificial regeneration moreover received considerable financial and logistical overseas' support. After independence former colonial interventions – through the supply of experts, policy directives, funding, etc. – were soon replaced by similar support from the Food and Agriculture Organisation of the UN (FAO), the World Bank and other bilateral funding projects, which played a considerable role in shaping policy and forestry practices in this period. Most of this was concentrated on training and the provision of equipment. Thus, the Forestry Department established at Ibadan University in 1963 was staffed and funded by the FAO, and when, in the 1970s, the government's Forestry Department was greatly expanded, the training of new staff was paid for by bilateral and multilateral aid agencies.[44] Much of the equipment provided was 'tied', i.e., the Deutsche Gesellschaft für Technische Zusammenarbeit (GTZ) projects provided German equipment, the Swedish International Development Cooperation Agency (SIDA) Swedish equipment. The World Bank also funded several large-scale projects, of which the most important was the Industrial Pulpwood Programme. Launched by the Nigerian government in 1978, its aim was to provide wood for the existing and newly planned pulp and paper mills. An additional World Bank loan secured in 1979 financed the establishment of 25,000 hectares of industrial plantation (EC-FAO 2003).

Funded by oil revenue and international support, tree planting took off in both northern and southern Nigeria, with around 20,000 hectares established annually in the 1970s and early 1980s (Parren 2003). The two main trees planted in these large-scale plantations were teak (*Tectonia grandis*) and gmelina (*Gmelina arborea*). Teak is a useful, versatile and popular timber, whilst gmelina was planted for pulp and paper production, as well as for timber. Deriving from South Asia, they are of the same family (*Verbenacea*), and silviculturally quite similar in that they are both medium-sized, fast-growing trees that thrive in tropical regions of both high and low rainfall (Lowe 1994: 14–17). Both, but especially teak, had become the 'tree of choice' worldwide, symbolising efficient industrial forest management, and were planted almost exclusively in tropical countries throughout the world in the 1970s (Amanor 2001).

In what was now Bendel State, a large number of gmelina and especially teak plantations were established in forest reserves cleared of existing forest cover by bulldozing. They usually occupied reserve outskirts in order to protect the forest inside (Lowe 1994: 40–41). One immediate effect of these plantations, therefore, was a significant reduction in biodiversity. In the longer term, of course, plantations were supposed to protect natural forests inside reserves by providing alternative sources of timber and could therefore, possibly, have contributed to maintaining biodiversity, as well as meeting timber, firewood and pulp demands. However, with plantations remaining prone to disease and fire, successful timber plantation management still required large amounts of labour; its costs made it highly dependent on continuous high levels of investment.

In the meantime logging continued to expand rapidly. Under the First Periodic Block (1945–1969), the UAC, or, as it is more commonly referred to locally, AT&P (African Timber & Plywood Company), had held 91.6 per cent of the concession in Benin Division, including all the concessions in Okomu Reserve (Adeyoju 1966: 158). When the Second Periodic Block (1970–1994) was launched, AT&P were again allocated the majority of concessions in Okomu, although new rivals such as the Italian plywood company Piedmont and smaller local sawmills were beginning to appear. The timber produced was now increasingly sold to the domestic market, especially when international timber prices crashed during the oil crisis in 1973 (Darling 1995: 14). Domestic demand rose so much that in 1976 a ban was put on the export of timber logs from Nigeria, but logging rates continued to rise, and allocations began to be made more frequently than working plans under the Second Periodic Block prescribed.

By the early 1980s the Benin area and Nigeria as a whole had been subject to over eighty years of colonial and post-colonial scientific forest management. Throughout this time there was an established, 'proper' forest department, with a cadre of foresters allocating logging licences, experimenting with regeneration methods, writing annual reports and participating in global forestry trends; generally seeking to 'systematically organise' Nigeria's forests on 'Indian lines', such as H.N. Thompson had promised in 1904. But, as I have shown here, this did not actually equate to successful, sustainable, consistent management, such as it is remembered today. For a start, scientific forestry had taken many decades to be fully established, with proper working plans becoming operative only in the 1950s. It had for long periods been riddled by financial constraints and staff shortages, and had undergone several sudden shifts in strategy, such as the non-application of the methods originally experimented with at Sapoba in the 1920s, the abandonment of the elaborate working plans of the 1930s, and the change from TSS to artificial regeneration in the 1960s. More significantly, even when working plans were fully implemented, forest management in Nigeria never achieved its actual aim, namely the creation of large-scale, sustainable timber production, where sustained high levels of timber removal are matched by constant regeneration. If for decades the department had struggled to encourage greater and more diverse logging, once logging did take off its expansion – due to technical and market factors rather than the activities of foresters – was so rapid that it was again largely beyond the control of the department. At the same time, even TSS, the most concerted effort at timber regeneration through natural regeneration, had disappointing results, whilst the more recently established plantations of exotic timbers involved large-scale forest clearance, were expensive to maintain and had not yet fully matured.

All this meant that, by the early 1980s, Benin forests and forestry were already in a difficult situation. Over the years a great number of standing large timber trees had been removed, but their successors were regenerating slowly and irregularly in forest conditions. By now, however, the timber industry was fully established and had become the most significant sector of the Benin economy, in ways that still enriched business and political elites disproportionately. Domestic demand for timber was growing rapidly, a demand which could not yet be met, if ever, through artificial plantations. Already then, working plans were difficult to maintain and logging allocations were made more frequently than they prescribed. These inherited difficult conditions played a significant role in the collapse of 'proper' forest management in recent decades.

Logging and Forestry in Edo State Today

Nigeria's economic and political crises of the 1980s and 1990s greatly affected forest management, bringing the investments in forest staff, equipment and plantations made possible through oil revenue in the 1970s to an abrupt end. Staff retrenchment in 1984 and again in 1997 significantly reduced forestry staff. As one of the forest officers in Benin City put it, 'where there used to be twenty forest guards, there are now only two; where there were a hundred silviculturists, there are now under ten, or even none at all'. As well as understaffed, the Forestry Department has been chronically under-resourced, and without vehicles, petrol and equipment the department's capacity to control illegal logging and maintain plantations has been severely restricted. The breaking down of paper mills in the 1980s meant many gmelina plantations became too overgrown for pulping (EC-FAO 2003), whilst teak plantations furnished a teak boom, with a large amount of teak exported to India for the production of garden furniture for European markets. Since then, however, plantations have been largely neglected, and there is now little teak left for export. Numerous former plantations situated along roads in Edo State are now only sparsely covered with small teak or gmelina trees and are frequently exposed to fire.

Real financial difficulties have been exacerbated by lack of political interest in forestry. Even though Nigeria signed up to the Tropical Forest Action Plan (TFAP) launched in 1985 and received UN funding, TFAP never went beyond the planning stage in Nigeria. All former working plans have been abandoned: foresters continued to make far more allocations than the Second Periodic Block (1970–1995) prescribed, and the planned Third Period Block (1995–2020) was never introduced. Instead, logging allocations are largely determined by personal connections or given to the highest bidder. The breakdown of working plans has been variously attributed to both civilian and military regimes. When Patrick Darling conducted research into logging allocations in Okomu Reserve in 1995, during Abacha's military reign, he was told that it was during the civilian regime of 1979–1983 that the best forests had been plundered, 'without the power of the gun'(Darling 1995: 14). In contrast, in the early 2000s when Obasanjo's civilian government was in power, forest officers told me that it was the military which 'destroyed all working plans' and just allocated everything out, because Edo State was seen as a 'gold mine'. It was indeed under the military regime of Abacha in 1995 that forestry was transferred from the Ministry for Agriculture and Natural Resources (MANR) to come under the direct control of the Governor's Office, which gave the Governor far more power to decide logging allocations.[45] However, whilst forestry was transferred back to MANR when civilian rule began in 1999, there has been no change in overall practices.

This last section, then, describes logging in Edo State today: the logging industry itself, much transformed since the 1970s, and the ways in which it is managed by the Forestry Department. It provides insights into the numerous illegal and semi-legal practices that have developed as well as the political economy and social networks shaping logging allocations and regulation, but also highlights nuances and resistance; not everyone participates in the same way.

Edo State's logging industry is by now highly fragmented. The multinational logging firms, in particular AT&P (UAC) which dominated the industry for so long, have disappeared. In Okomu Reserve, AT&P's almost exclusive hold on allocations came to an end in the 1980s and 1990s, when an increasing number of allocations were given to local contractors, including the Iyayi Group and several Udo chiefs. AT&P and Piedmont ceased their logging operations in Edo State altogether in 1995. According to loggers today, prices were systematically undercut by the Western Metal Products Company Ltd (WEMPCO), a Hong Kong-based metal processing firm that branched out into logging and plywood production in the 1990s. At the same time, changing exchange rates and a diminishing number of trees combined to make logging and plywood production in Nigeria increasingly uneconomical for AT&P. The next largest company, the Iyayi Group, continued logging in Okomu Reserve a few years longer, but recently only small operators have been logging in the reserve. Amongst these are the family of the Chief of Iguowan and various other Udo chiefs.

If not in Okomu, there are still a few larger operators in Edo State, including the Iyayi Group. Caesar Iyayi, the son of Efionayi Iyayi, owns a large sawmill in Okada, which buys wood from a number of suppliers in addition to its own large fleet of trucks. Managed by Italians, it primarily produces parquet flooring for export to Europe, made from apa (*Afzelia spp.*), obeche and other timbers. Other larger operators have fleets of twenty or so trucks and log in many parts of Edo State and beyond; in recent years, several have started going west into Ondo State and other parts of southern Nigeria. Anticipating that there are only a few years of profitable logging left, many have begun to diversify into other business interests by investing timber money in hotels or petrol stations and reducing their timber operations. In addition, there are a number of medium-sized firms, which combine fleets of a dozen or so lorries with one or two sawmills; others have only one or two trucks as well as a sawmill.

However, many people in the Edo timber industry today operate on a considerably smaller scale. Often, small group of villagers hire chainsaws and cut trees in the forest, and then sell these in situ to the 'Steyr' people, who may have hired a truck rather than own it themselves. These sell the logs to a different set of the people doing 'conversion', who take the logs to the sawmills and later sell on the planks. Amongst these are several prominent 'big' women, who have built up very successful timber conversion businesses. In practice, there are a number of different kinds of arrangements and complicated structures of borrowing and advancing money between the different parties. One sawmill I visited, for example, was owned by a woman whose husband was a timber contractor. He had a fleet of six Steyr trucks, of which three were in good working order just then. One machine was operating and cutting obeche brought in by a woman working in conversion; it took rather long as the machine broke down frequently and was running on a generator because municipal electricity was unavailable. Overall, there is today a considerable divergence between well resourced, large operators who command operations from cutting to the selling of planks, and increasingly small operators who are often struggling to survive – 'it takes money to make money', as one interviewee put it.[46]

The multiplicity of operators is mirrored in the different practices that have developed in the interaction between loggers and forest staff. As previously discussed, working plans have been abandoned and the planned Third Periodic Block (1995–2020) was never introduced. Instead, allocations are now done, as one Forest Officer put it, 'by hunch'. In order to log in a particular forest compartment, logging firms still have to apply for licences, along more or less the same lines first set out in 1899 (Forestry Department 2002a). Allocations are made every three years. Thus, allocations were running out in December 2002, and new ones were being prepared in January 2003 and again in 2006 when I revisited. In the periods between the expiry of old licenses and the beginning of new ones, all logging is officially banned. Licences in 2003 cost ₦350,000 (£1,875) in the first year, then ₦150,000 in the second, if the contractor wished to continue. This can also be paid by someone else, who might wish to go for 'relics' – trees that the first contractor left standing. Officially there are still regulations about the minimum girth sizes of trees being cut (currently four feet), and all trees cut are supposed to be inspected by the Forestry Department. Nowadays this does not happen in the forest itself but at roadblocks as logs are being transported out. Logs coming out of Okomu Forest, for example, are inspected by forest guards stationed at Udo.

The official process therefore still broadly follows long-established bureaucratic procedures, including details such as the marking of logs with a 'hammer' by foresters once they have been inspected;[47] a continuity in practices similar to that noted by Anders (2004) in his study of civil servants in Malawi. Forestry also still generates some, albeit falling, official revenue for Edo State; in 2001 this was ₦28 million (£150,000).[48] However, logging allocation itself is no longer determined by considerations of forest management or tree regeneration. Logging licenses being allocated 'by hunch' means that larger companies, with better political connections and more resources, usually manage to use these to pay forest officers additional charges and have the best compartments allocated to them. It was reputedly because Efionayi Iyayi had lent money to the Governor that the Iyayi Group obtained its logging concession in Okomu Reserve, which it later used to claim the land and start a rubber plantation. And still today, with Caesar Iyayi a prominent 'chieftain' of the People's Democratic Party (PDP), Nigeria's dominant ruling party, the Iyayi Group gets some of the best allocations in Edo State and beyond. There is, here, a clear continuity in the patterns that emerged in the 1930s, when most of the additional allocations for local contractors that Oba

Akenzua II secured through persistent campaigning were given to close political allies. Thereafter, local allocations went primarily to members of the economic and political elite, not least to the *oba* himself.

A number of further practices have recently developed in addition to regular allocations of logging licenses. These include the rather less official system of 'seize and release', whereby timber contractors first go into an area and cut timber, inform the Forestry Department afterwards and then pay a fine. Here, timber contractors may end up paying ₦150,000 (i.e., less than the official price), but much of this may go to forest staff directly and contribute rather less to revenue collection. There is also a considerable amount of illegal logging of different kinds, as there has been for many decades. The word 'illegal' is generally applied to very small operators who cannot afford a licence and try and log without one. A number of such small illegal operators logged in Okomu Reserve, especially around Ugbor and Ofunama in the south of the reserve where their operations are organised by local chiefs. The Chief of Iguowan, the small village just north of the national park, began logging in Iguowan Lake during the time I was there in 2001–2002, the very area that had previously been protected and rescued from ORREL's expansion on the grounds that it was a sacred grove and of local significance.

Several informants pointed out that small illegal operations were risky and expensive – when caught both fines and bribes could be quite high. That this was not always the case is illustrated by an encounter between a forest officer and illegal loggers I witnessed. When the Conservator of Forests of Ovia South-West LGA took me and an Iguobazuwa chief into Iguobazuwa Reserve in January 2003 – a rare excursion away from the office – we came across illegal loggers inside the reserve. After taking the keys of their Steyr vehicle, the Conservator attempted to start the engine to drive it away, as vehicle confiscation was the most effective way of arresting loggers and getting a fine out of them, especially if they have hired the truck by the day. The loggers responded by kneeling in front of him and imploring him not to drive away the vehicle. They also 'settled' the Iguobazuwa chief by giving him some money (behind one of the opened truck doors so that I would not witness this transaction); as one of the traditional rulers of Iguobazuwa he also had some claims on forest revenue. In this way, possibly promising further payments the following day, they eventually managed to persuade the Conservator to return their keys. All this was done in quite a relaxed way, with loud R&B music blasting from the truck throughout the entire episode. Nevertheless, the general perception was that operating as a small-time, under-resourced, illegal logger involved considerable risks and was not always profitable.

There are also a whole range of illegal practices amongst operators who do obtain official licences. Thus, for instance, a common tactic is to

Figure 12 Typical log transport on an old steyr vehicle, showing small log sizes today. Photograph taken by the author, 20 August 2006.

obtain a licence for one compartment but then to use this to also log in all the compartments surrounding it without a licence; Iyayi, for example, was once found logging thirty miles away from his actual concession in Okomu Reserve. Loggers operating in Okomu Reserve also often transport logs out at night, in order to avoid paying government fees at the Forestry roadblocks in Udo, where their logs are supposed to be stamped by forest guards. 'Fees' here may well have included additional payments for logs under the minimum girth size. Most of the logs I observed on trucks transported along roads or at sawmills were considerably smaller than four feet; according to one logger average girth size was then about 2.5 feet (Fig. 12). Night transport is also a common practice during the month when there is an outright ban on logging. In addition, larger operators can also come to special arrangements with forest officials to circumvent the law. I accompanied one of these, Duke, into Ondo State, where he was setting up a new logging operation. He showed me some letters he had obtained from the Ondo State Forestry Department. These gave approval to collect illegally felled logs from Ifon Reserve, which is officially a game reserve in which all logging is banned. They did not mention who had cut timber there illegally – the logging contractor himself, as everyone knew.[49] Overall, whilst forest officers make sharp distinctions between legal and illegal loggers – blaming illegal loggers almost exclusively for over-exploitation and forestry problems today – these distinctions are rather blurred in practice; no single allocation or logging contractor operates in a strictly 'legal' way.

Indeed, for all loggers the creation and maintenance of personal relations and 'friendships' with forestry staff and politicians is crucial, as in dealings between state officials and the public in many other parts of Africa (Blundo and Olivier de Sardan 2006; Olivier de Sardan 1999). This is the case in one-off encounters between small-scale illegal loggers and forest guards, such as the one in Iguobazuwa Reserve described above, but even more so for operators wishing to log, legally or illegally, longer term and on a larger scale. Loggers routinely use strategic gifts to build up relations; at the end of a visit to one of Benin City's sawmills, its owner gave a large present to the forest officer who had taken me there, and small amounts of money were frequently exchanged with visitors to the Forestry Department. Offices of the Iyayi Group displayed well-wishing cards by forestry staff, whilst the forest office had Iyayi's and other timber companies' calendars hanging on its walls. Forestry offices both in Benin City and in Iguobazuwa were always busy as visitors continuously dropped in with requests, bringing gifts or just 'greeting', all to build up and maintain good relations. There are also direct connections, with several former forestry staff made redundant in the last big phase of retrenchment and now working as timber contractors. Establishing such relations paid off. Duke, for example, claimed he was able to transport logs from Ondo to Edo State, which is otherwise not allowed, because he was a friend of the Edo State MANR Commissioner. Importantly, his success here was not purely the product of actual bribery but also involved a number of other factors. Thus, Duke also proudly told me that he only spent about 40 per cent of the amount that others had to spend on 'PR', as he had many friends in high places. As Duke's boasting demonstrates, the importance of social relations means that well-positioned logging contractors are able to play the system in a different way to small chainsaw operators, who at times have to pay such high fines or bribes that they risk losing rather than making money. Even if small- and large-scale loggers are all engaged in illegal practices and seek to foster 'friendships', their methods and experiences are very different, just as those of small cocoa and plantain farmers in Okomu Reserve diverge from those of politicians officially being given large amounts of reserve land.

Similarly, it is important to emphasise that not all forestry staff participate in arrangements with loggers in the same way. Timber operators I spoke to frequently compared Edo officials unfavourably to those of Ondo State, saying that those in Edo State were more 'lax'. In Ondo State, they claimed, the process could be highly cumbersome, with negotiations and 'PR' payments at every stage, whereas in Edo State loggers sometimes only had to pay the forest guards and could get away with more. In support of this I was told that one forest director in Ondo State had genuinely been trying to stop logging in Ifon Game Reserve, but once the Commissioner and everyone else had condoned it, there was no point in his 'one-man crusade'. Such comparisons, of course, may reflect individual experiences and perceptions as much as actual differences. Certainly, Edo forest officers are subject to their superiors and political leaders in the same way as their colleagues in Ondo State. Thus, they themselves frequently claimed that they wished to combat illegal logging far more effectively, but that when they caught illegal loggers they often received phone calls from politicians asking for their release, 'so our hands are tied'. Nevertheless, it seems likely that different historical trajectories mean that, overall, practices do vary between different state forestry departments. At the same time, individual foresters bring different attitudes to their job – in Edo State as much as in Ondo State. In Ondo I witnessed a conversation between forest guards talking about one old guard who kept on writing down the actual girth size he measured; everyone tried to tell him that that was not how it was done, but he was too old to change. At the Ovia South-West LGA office in Iguobazuwa, meanwhile, one young forest guard I met was highly conscientious and enthusiastic, making a rather different impression than his generally bored and jaded-looking colleagues. Officers at the Forestry Department in Benin City also differ in their concerns and temperaments. Some, like the Wildlife Officer, talked to me at great length about their education and training, their commitment to forestry and conservation and their frustrations with the current state of affairs, whilst others were quite open about their own involvement and the profitability of their positions; these, indeed, are the ones with frequent visitors and a more developed social life. Even these officers, however, displayed enthusiasm for forestry and recalled their training and early work years with nostalgia. Edo foresters, like bureaucrats elsewhere, are not simply all corrupt officials; rather, they draw on a multiplicity of legal and moral frameworks (Anders 2004; Blundo and Olivier de Sardan 2006).

Rural communities' attitudes to recent developments are also ambivalent. Some members of local communities, such as the chiefs of Iguowan, Ugbor and Ofunama in Okomu Reserve, are often themselves involved in illegal logging, as also observed in Cross River State (Blackett and Gardette 2008). But there is also much resentment as well as active resistance against loggers. In Edo State this is not as organised as in Cross River State, where women groups and other local organisations successfully campaigned against WEMPCO's logging activities and the pollution caused by its sawmill, which eventually led to the closure of the mill in 2004.[50] Nevertheless, there are instances of community resistance in Edo State, too. At Iguesogba in Ovia North-East LGA, for example, loggers were banned entirely for some time in order to allow tree regeneration.[51]

Admittedly, it was difficult to uphold this ban, especially once someone from the community 'had shown them the way', as one inhabitant of Iguesogba put it. Local people mobilising against illegal logging can moreover also be seeking payments from loggers, rather than forest protection as such. Thus, at Udo Forest Reserve in Igueben LGA, in northern Edo State, local people had tried to prevent loggers from entering by placing a series of small logs across the road going into the reserve. They were relatively easy to remove during my own visit to the reserve, which suggests that they were part of a (then unmanned) roadblock rather than prohibiting entrance altogether. Nevertheless, even without active resistance local people often resent loggers. At Iguowan, for example, there were frequent complaints against the chief for starting to log within Iguowan Lake, which Iguowan inhabitants had actively protected from ORREL's expansion. I was talking to a group of female farmers in Iguowan in early 2002 when we saw a large tree in the lake being felled; all the women wailed loudly as they observed this. Moreover, whilst government plantations and regeneration programmes are all but abandoned, some farmers in rural areas have themselves started to plant timber trees on their farms, or are fostering those that have germinated naturally. The *odionwere* of Okapha II (north of Udo), for example, grows white and black afara in some of his farms, which he rotate every twelve years, and some farmers in Udo have also started planting trees on their land. Indeed, the Conservator of Forests in Iguobazuwa told me – with some indignation – that the communities were managing to protect their own land, with a large number of young saplings coming up there, whilst he was not able to do the same in reserves.

Overall, then, government forest management has indeed largely collapsed and corruption and patronage politics are rife: instead of scientific working plans, official logging allocations are now made largely 'by hunch', on the basis of political connections, and an additional plethora of semi-legal and illegal arrangements has developed between forest staff and loggers. Such practices are not entirely new, however; both the allocation of compartments to political allies and illegal logging have been part of the logging industry for some time. Moreover, the situation today is more nuanced and complex than its standard portrayals convey. Just as large- and small-scale loggers operate quite differently, so forestry staff are not all simply uniformly corrupt but are, to varying degrees, still motivated by their forestry training. And whilst some members of rural communities are themselves involved in logging operations, others seek to regulate it themselves and are regenerating timber trees on their own land.

More importantly, the current situation needs to be understood in its longer term context. I have stressed above that in the late 1970s and early 1980s, before Nigeria's economic crises and the decline of forest management, the system of forest management was already creaking, with domestic timber demand alone fast outstripping the volume of supply available under the Second Periodic Block. Perhaps under different circumstances, with ongoing high expenditure and political commitment, a successful transition to a timber economy based largely on teak plantations might have been possible. In Omo Reserve in Ondo State, for example, there are plantations of tall gmelina and teak trees lining the Benin–Lagos road, which are managed by WEMPCO; perhaps with similar investment government plantations, too, could still be producing teak and gmelina. However, considering the ecological, economic and political challenges the continuing creation and maintenance of large-scale plantations would have involved (in terms of disease and fire control, loss of biodiversity, costs, labour and land politics), this hypothetical scenario seems unlikely even under the most favourable circumstances. Given the growing demand for timber, it would similarly have been extremely difficult to maintain the rotation system of the Second Periodic Block and to start with the Third Periodic Block as originally envisioned in 1945, even with a well funded and highly committed forestry cadre in charge. Clearly, some form of revision was becoming inevitable by the 1990s. Instead of formally revising working plans, however, they were simply quietly abandoned, similar to how reserve land was being made available.

In contrast to land conversion, the social and ecological outcomes of recent developments in timber extraction appear to be more or less as described in standard accounts. Whilst the logging industry continued to thrive, providing employment and income to thousands of workers and timber to a growing domestic market, it seems that timber supplies in Edo State really were getting low in the early 2000s. With average girth sizes already small, loggers were trying to get as much out of timber as possible, before Edo State was completely 'finished'. In 2002 loggers began to cut *Ceiba* trees (*Ceiba pentandra*, also known as silk cotton), which throughout the twentieth century had been left standing because their softwood was not very useful. Today, however, it can be used for making (rather inferior) plywood with glue, and recently many people started 'rushing' to *Ceiba* trees. Several of these large and conspicuous trees near Okomu Reserve were felled in late 2002, despite the fact that profit margins were low – 'desperation work', as some of the larger operators dismissively called it (Fig. 13). Indeed, as described above, loggers have started to operate in neighbouring states and many of Benin's 240 or so sawmills are supplied with timber from beyond Edo State. As logging profits are falling, several larger operators have also begun to diversify into other businesses such as petrol stations and real estate. However, it must again be stressed that after decades of logging and little timber regeneration, Edo forests were already significantly depleted of their timber stocks by the 1970s. In this respect

Figure 13 Recently cut *Ceiba* tree near Iguowan. Photograph taken by the author, 18 December 2002.

too, recent decades present not a new era of reckless logging after decades of sustainable management, but the latest stage of an ongoing process.

Conclusion

To any recent observer of forestry in Edo State, the disarray of logging regulations, the lack of regeneration programmes and the depletion of the state's timber trees have all been so evident, and the links between them so easy to draw that it seems obvious that forestry's main problems today are lack of political interest, rampant corruption and widespread illegal logging. This chapter has probed this analysis in two ways. For one, it has described contemporary logging and regulatory practices, formal and informal. In this way, it has provided a better sense of the manifold manifestations of corruption and patronage networks in the logging sector, but also of their limits: not all foresters participate in corrupt deals, whilst rural communities have developed strategies to combat illegal logging and to regenerate timber themselves. The chapter's main argument, however, has been historical. For one, it has shown that current practices, such as illegal logging or the allocation of logging concessions to political allies are not new but have a long history in the Benin area: from the beginning, timber royalties were central to the ways in which colonial income opportunities bolstered the patrimonial Edo political economy. Most importantly, the chapter has shown that, for a number of reasons, scientific forest management never achieved its central goal of creating a sustainable timber industry. Forestry suffered not only from financial shortages and setbacks that for many years prevented the full application of working plans. Even when this was finally achieved in the 1950s, it still did not have the results hoped for. Scientific forestry was based on a fundamental belief in government control over both forests and logging activities, yet it never achieved either: forest regeneration remained problematic throughout the twentieth century, and logging levels and demand for timber were always determined by technological and other external factors rather than by interventions by the Forest Department. By the late 1970s many of the large timber trees once found in Benin forests had already been removed. A series of attempts at different methods of natural and artificial forest regeneration had largely failed, whilst timber demand had grown beyond what could be met through the allocations prescribed under the Second Periodic Block. Putting Edo forestry's current situation in this longer term context shows that many of its current problems result from scientific forestry's own legacies, rather than from its recent abandonment.

Notes

1. National Archives (NA), CO 879/69, Enclosure 3 in No. 138, 177, Address by the Conservator of Forests, Southern Nigeria, before the Chamber of Commerce, Liverpool, 16 September 1904.
2. Moor drew up a *Memorandum of proposed Regulations with regard to the carrying on of Lumber work*, which he sent to the Secretary of State. In the accompanying letter, he wrote confidently that 'the safeguarding of this industry is ..., to a great extent, a matter of practical supervision and control and as the areas where it can be carried on in these Territories are at present to an extent limited owing to the question of transport it does not seem to me that there should be any insurmountable or great difficulty in establishing effective control'. NA, CO 444/3, No. 201, Sir Ralph Moor to the Secretary of State, 26 November 1899.
3. West African mahoganies were not actually true, but 'false' mahoganies (see Melville 1936). Several hardwood species from West Africa were referred to as 'mahogany' at the time, but in the Benin area mahogany usually meant *Khaya ivorensis*, or lagos wood.
4. NAE, CSO 3/5/3, Sir Ralph Moor, HB.M Commissioner and Consul General, to the Secretary of State for Foreign Affairs, 14 November 1898.
5. NA, CO 444/3, No. 201. Memorandum of proposed regulations with regard to the carrying out of lumber work. Enclosure in a letter by Sir Ralph Moor to the Secretary of State, 26 November 1899.
6. Ibid.
7. Ibid.
8. Ibid.
9. Girth sizes were measured at the height of cutting, six feet from the ground.
10. *Civil Service Handbook* 1907.

11. The introduction of tramways features prominently in the annual reports of the time. For further descriptions, see Adeyoju (1966) and Egboh (1985).
12. Out of a total of 144 areas leased in that year, only 51 were worked. 'In the case of Miller Brothers Limited, only 4 areas out of the 26 granted under 1901 rules were worked, but this is because most of the areas of 9 square miles are exhausted' (McLeod 1908: 245–46).
13. Calculated from a table of forest export statistics, Appendix II, Unwin (1920).
14. The earliest mentioning of a local timber trader I found is from 1903. See Nigerian National Archives Ibadan (NAI), CSO 16/5/55, No. 131/1903, Report on a Visit to Siluko. Dumett (2001) also describes some indigenous timber traders in Ghana.
15. Timber trees were divided into different classes, according to their economic importance: first class trees included *Khaya ivorensis*, *Entandrophragma spp.* and *Milicia excelsa* with an eleven-foot minimum girth limit; *Gaurea spp.* and *Lovoa klaineana* with a ten-foot minimum, *Sarcocephalus esculentus* with an eight-foot minimum, *Funtumia elastica* and *Ficus Vogelii* with one of six feet, and *Elaeis guineensis*, the oil palm, without any girth limit (Unwin 1920: 228).
16. Nigerian National Archives Enugu (NAE), Legislative Council Papers 1912, No. 35, Annual Report on the Central Province for the Year 1911, 5.
17. Some mahogany plantations planted in the Okomu area appear to have generated many tall mahogany trees, which were cut sixty years later, according to the Chief of Ato (Igbene).
18. Thompson and other colonial foresters of his generation had been trained at the Royal Indian Engineering College at Coopers Hill.
19. The Benin Native Administration Forest Circle was an administrative division used within forestry only. This area covered most of but not the entire Benin Division.
20. NAE, Sessional Papers, No. 7, 1934, Report on the Commercial Possibilities and Development of the Forests of Southern Nigeria, by Major Oliphant; NAI, BEN DIST 4, 27354, Vol. I, Conservator of Forests to Chief Secretary to the Government, Lagos, 21 March 1935.
21. Ibid.
22. The Benin Circle is the Benin Native Administration Forest Circle. The title of the Conservator of Forests of the Benin Native Administration Forest Circle was usually abbreviated to Benin Circle Conservator of Forests.
23. NAI, IB FOR DEP 1/760, 223/WP, 1697D, Annual Report of the Benin Native Administration Forest Circle 1935; Conservator of Forests to Chief Secretary to the Government, 21 March 1935.
24. NAI, IB FOR DEP 1/760, 223 WP, 1697 D, Annual Report of the Benin Native Administration Forest Circle 1935; IB FOR DEP/120, Annual Report of the Silviculturist of the Forest Department 1934.
25. NAI, IB FOR DEP 1/760, 223 WP, 1697 D, Annual Report of the Benin Native Administration Forest Circle 1935, 10.
26. Conservator Ainslie had already written in 1929: 'it will be necessary as each of the concessions agreements expire to insist on new conditions which will necessitate the felling of all marked trees on the area whether of presently exploitable species or not, the trees having been previously marked by the Forestry Officer in direct charge of the area. Only by this means can the permanency of the mahogany trade be assured' (Ainslie 1930: 6).
27. NAE, CAL PROF 53/1/558, A further report on forestry development in Nigeria by J.N. Oliphant, 10.
28. NAI, IB FOR DEP 1/760, 223 WP, 1697 D, Annual Report of the Benin Native Administration Forest Circle 1935, 5.

29. NAI, IB FOR DEP 1, 1897F, Handing Over Notes of the Benin Native Administration Forest Circle, by R.F. Clarke-Butler-Cole, Assistant Conservator of Forests, 6 June 1941, 12.
30. See also NAI, IB FOR DEP 1, 1897F, Handing Over Notes of the Benin Administration Forest Circle, by R.F. Clarke-Butler-Cole, 1942, 33, and BEN PROF 1, BP 640, Report on the Forest Activities of the Benin Native Administration, 27 October 1933, 3.
31. On the rise of these species during this time, see Adeyoju (1966).
32. NAI, BEN DIV 6, BD27/VI, Annual Report of Benin Division 1938, 13.
33. NAI, 1B FOR DEP 1/889, 1897J, BC 206/5, R.F. Butler-Cole, Conservator of Forests in Benin Native Administration to Oliphant, Chief Conservator of Forests, 26 January 1940.
34. NAI, IB FOR DEP 1, 1897F, Handing Over Notes of the Benin Administration Forest Circle, by R.F. Clarke-Butler-Cole, 1942.
35. Ibid., 8.
36. Interview with Donald McNeil, former Director of AT&P, September 2003.
37. NAI, BEN DIV 6, BD 27/Vol. XIV, Annual Report on the Benin Division 1947, 9.
38. NAI BEN DIV 6, BD 27/Vol. XV, Annual Report on the Benin Division 1948; Rosevear (1952a, 1953).
39. NAI, BEN DIV 6, BD 27/Vol. XIV, Annual Report on the Benin Division 1947, 10.
40. Conditions for labourers in forest camps were often harsh, with workers having to walk sometimes four or more miles to fetch water. NAI, BD 27/ Vol. XV, Annual Report of the Benin Native Administration Forest Circle 1948, 8.
41. Interview with Donald MacNeil, September 2003.
42. NA, CO 879/69, Enclosure 3 in No. 138, 177, Address by the Conservator of Forests, Southern Nigeria, before the Chamber of Commerce, Liverpool, 16 September 1904.
43. In 1963 'growing doubts as to the efficiency of TSS as practised in the Western Region, coupled with doubts as to the justification of spending vast sums of public money on an uncertainty resulted in the Chief Conservator of Forests, at the beginning of the year, ruling that pre-exploitation operations should only be carried out in specially approved areas, and, in most charges, abandoned altogether.' Institute of Commonwealth Studies (ICWS), Annual Report of the Western Region, 1962–1963, 7.
44. Lowe (1994) and interview with David Ward, formerly at the Ibadan Forestry Department, Forres, July 2003.
45. Moreover, Darling points out that 1982–1992 receipts for timber concessions in Okomu Reserve revealed clearly that under the military regime allocations were made almost exclusively from parts of the reserve that, under the working plan, were supposed to be regenerating (Darling 1995:14).
46. Interview with Alfred Ohenhen, Iguowan, 18 December 2002.
47. Interview with forester about procedures today, Benin City, 16 August 2006.
48. Revenue collected between 1991 and 2002 ranged from just under ₦2 million in 1991 to over ₦70 million in 1997. However, the Naira's worsening exchange rates need to be taken into account.
49. In general, Ondo's richer timber resources had recently attracted many loggers from Edo State, who, often armed, entered Ondo forests without a license in order to cut timber, especially apa.
50. http://www.earthisland.org/journal/index.php/eij/article/west_africa_rainforest_network/, accessed 17 February 2010. See also Johnson (2003).
51. I visited Iguesogba on 27 November 2002.

CHAPTER 4

Reinventing Farm and Forest
The Changing Forms of Taungya Farming

Introduction

Taungya farming, an agro-forestry system originating in Burma, was introduced to Nigeria by the colonial Forest Department in the late 1920s. Meeting both agricultural and forestry needs, agro-forestry has attracted particular interest since the rise in concern over tropical deforestation in the 1980s, but it was already a widely implemented policy throughout the tropical world in the colonial period. In the Benin Division much hope was put into Taungya farming, which promised to be both a cheap form of afforestation and a means to provide farmers access to good farmland. For some decades Taungya was indeed quite successful; by the 1960s it had become a key afforestation method. Today, however, whilst large parts of reserve land continue to be allocated annually to farming under the Taungya system, tree planting has largely ceased. It is generally accepted that Taungya has 'failed woefully' as a form of afforestation, and that instead it has now become 'a means of destroying the forest'(Forestry Department 2002a; Lowe 2000). Like the conversion of reserve land in general and the collapse of former logging regulations, Taungya's decline is explained through lack of investment, mismanagement and corruption.

This chapter again offers an alternative interpretation to the dominant narratives of Taungya's decline. Its approach, however, differs from the preceding two chapters, in that its emphasis is less on a re-evaluation of past management practices than on Taungya's recent transformation. Taungya's early history is interesting as it demonstrates colonial foresters' changing perceptions of farm and forest relations whilst also epitomising scientific forestry's fundamental need to organise and control these relations. However, here the discrepancies between desired and actual outcomes were less than in previous chapters, which is why it will be discussed relatively briefly. Focussing instead on the more recent period, this chapter looks, first, at both foresters' and farmers' explanatory narratives of decline. Such competing narratives also exist, of course, in relation to forest conversion and the collapse of working plans and have indeed been briefly touched upon in previous chapters. Those relating to Taungya are discussed in more detail, however, because it is the very decline and failure they seek to explain that is challenged here. Rather than having 'failed woefully', this chapter demonstrates that Taungya in its present form fulfils an important function in providing access to land in the Okomu area, and, moreover, need not necessarily be seen as a means of destroying the forest. Alternative interpretations of its ecological effects emerge in particular when Taungya farming is put into the larger context of local land-management practices, on unreserved as well as reserved land. This re-interpretation of the transformation of Taungya therefore offers a further, important angle to the book's overall challenge of existing understandings of the unfolding of forest management in Edo State. This being the last of the three chapters examining one forest policy over time, it concludes with a summary discussion of the findings of all three, drawing together the different ways they have probed the dominant understanding of the collapse of forest management.

Taungya in the Benin Division: Controlling Shifting Cultivation

Up to the 1930s colonial foresters in Nigeria routinely described shifting cultivation as a 'pernicious' system of cultivation that was destroying forests on an 'immense scale'. As discussed before, such damning portrayals were made particularly in the course of calls for further reservation, which needed to be justified as a measure of forest protection. In reality, however, foresters had already begun to view it more favourably from the 1920s onwards. In 1922 Thompson acknowledged that local farming practices appeared, after all, to be 'the most suitable under the circumstances' and that they were not as destructive as previously assumed (Thompson 1923: 11). Even more than in Nigeria, in other colonies foresters also recognised that shifting cultivation could potentially play an important role in tree regeneration. The following resolution was passed at the British Empire Forestry Conference held in Canada in 1923, which Thompson quoted in his 1924 Annual Report:

> The practice of shifting cultivation, except when controlled as an integral part of Forest management is a serious menace to the future of certain portions of the Empire. At the same time this Conference recognises that, if strictly controlled shifting cultivation may, under certain conditions, be made to serve a useful and even necessary purpose in Sylvicultural operations,

particularly in connection with the formation of plantations. When applied to such useful ends this Conference favours its encouragement under control. (Thompson 1925: 11)

This passage is interesting in that, on the one hand, it shows that foresters recognised the ecological virtues of shifting cultivation, advocating its integration into silvicultural operations. We saw in the previous chapter that in Nigeria too, foresters over time gained more and more insight into the beneficial effects of farming on timber regeneration, realising that many light-demanding timber species regenerated poorly within high forest but well once forests had been opened up through farming. Yet on the other hand, the above passage also clearly conveys the limits of this recognition. It was inconceivable that farmers might be allowed to continue to farm in the forest as they had before, where and when they wanted. For one, if brought to its logical conclusion this strategy would have potentially made their own contribution as professional foresters redundant, which was not in their interests. At the same time, the passage conveys foresters' profound unease with the seeming chaos of local shifting cultivation practices. For both these reasons, shifting cultivation could only be encouraged when 'strictly controlled', 'as an integral part of Forest management'. Therefore, rather than leaving farming patterns and tree regeneration to chance, farming activities and tree regeneration had to be strictly controlled by scientific foresters. This approach bears striking resemblance to that in northern Nigeria, where local dryland forestry practices were recognised but consistently devalued and deemed to be in need of government regulation by colonial forest officers (Cline-Cole 1996).

In order to arrive at a suitable Taungya system for southern Nigerian conditions, the Forest Department included agro-forestry trials as part of its research into natural and artificial regeneration methods at Sapoba Research Station. The first of these trials, conducted by St. Barbe Baker in 1927, were referred to as Chena Plantations, an agro-forestry system that St. Barbe Baker had experience with in Kenya. At the same time, J.D. Kennedy was sent to Burma as part of his training, 'in order to find some inspiration for solving afforestation problems in Nigeria'. He was particularly impressed by the Taungya system practised there, and applied his insights to the agro-forestry trials at Sapoba when he took over the station in 1928 (Kennedy 1930). In Burma Taungya originated as a local agro-forestry system that had been adapted in various confrontations between farmers and the British Forestry Service (Bryant 1994). By the time it was introduced to Nigeria and other colonies in Africa, such as Tanganyika (Sunseri 2009), it had become an integral part of scientific forestry.

Under the Chena/Taungya system,[1] twenty-six acres of land at Sapoba were cleared by local farmers and departmental workers. Farmers then planted yam, maize, groundnut, okra, cotton, pumpkin, pepper and melon, as well as plantain – because of their rate of growth and heavy shade, plantains were regarded as harmful to tree seedlings but had to be tolerated as a staple part of people's diet; they were planted around the boundaries (Kennedy 1930: 224). In addition to food crops they planted a number of indigenous hardwood seedlings, including mahogany (*Khaya ivorensis*), sapele wood (*Entandrophragma cylindricum*) and obobo (*Guarea spp.*). In the early twentieth century colonial foresters still spent considerable effort on attempting the regeneration of slow-growing indigenous trees in plantations, in Nigeria as well as East Africa (Schabel 1990); as described in the last chapter, it was only several decades later that Nigerian forestry switched to fast-growing exotic trees. In addition, Kennedy noted that 'curiously enough ... a considerable quantity of naturally sown iroko (*Milicia excelsa*)' seedlings also emerged and thrived in between the field crops. Iroko was, of course, another tree that had always grown particularly well on abandoned farmland (see Chapter 2). By 1928 Kennedy reported that the Chena plantations 'bid fair to become an outstanding success' (Ainslie 1929: 15). The following year he further reported that 'the farmers are beginning to understand what is required of them, and realise the benefits of the system to themselves, and a good healthy rivalry exists as to whose is the best farm!' (Ainslie 1930: 14). The scheme does indeed appear to have been popular, as several farmers from outside Sapoba Reserve asked to participate. The Director of Forests welcomed this, 'for the more this scheme is advertised the better and the sooner will it be realised how utterly wrong and inimical indeed to the whole rural economy of the country is the view that farming interests and those of forestry are necessarily mutually antagonistic' (ibid.). Indeed, Taungya was then widely perceived as a 'heaven-sent solution to the problem of forestry versus agriculture in the tropics' (Rosevear 1954: 17).

These early experiments with Chena/Taungya farming were deemed so successful that by 1935, Taungya was predicted to be 'of increasing importance as a method of regenerating poorly stocked or secondary forest'.[2] Indeed, it was to form an 'essential part' of the Benin Forest Scheme. Oba Akenzua II agreed that in return for granting the expansion of forest reserves by an additional 1,000 square miles, 1,000 acres annually would be regenerated through Taungya.[3] Under the scheme the Forest Department would choose a suitable area of degraded reserve forest near villages with land shortage and would allocate about two acres to each participant. In contrast to local farming practices, under which farmers tended to leave a number of trees standing, under Taungya farmers were instructed to clear-fell and burn the plot completely, so that no previous cover remained at all. They then had to plant their food crops, whilst silvicultural forest staff was responsible for planting and maintaining seedlings.[4] The seedlings at this time, chosen for their economic value, were again mainly hardwoods, such as mahogany (*Khaya ivorensis*) and sapele wood (*Entandrophragma cylindrica*), and were planted about sixteen feet apart. Farming would cease after one or two years, at which stage clearing and thinning of undesirable regrowth began. Rotation periods were envisioned to be around seventy (Clarke-Butler-Cole 1943), or even a hundred years.[5] Overall Taungya seemed to be working out quite well, as farmers were 'quick to recognise the advantage of being given each year a portion of good forest land instead of the over-cultivated and impoverished land to which they were formerly restricted', and, it seems, willingly cooperated (ibid.: 104).

The acreage planted annually in this manner increased steadily but slowly. By 1948 still less than the envisioned 1,000 acres were planted annually and in total there were still only 6,575 acres (just over ten square miles) under Taungya in the Benin Division.[6] Some first disappointments were noted by then too, with foresters in charge finding Taungya farms in Benin poor by Burma standards and their cost 'out of all proportion to their probable value at maturity'.[7] But on the whole, in the Benin Division Taungya was deemed a success as farmers here, in contrast to other parts of Nigeria, showed remarkable enthusiasm for the scheme. This may well have been because large-scale reservation had already created significant shortages of community land here, and Taungya presented a welcome opportunity to gain access to good forest land.

In the 1950s the Forest Department concentrated its activities on the Tropical Shelterwood System (TSS), which was seen as a more cost efficient form of timber regeneration, so Taungya planting continued but was not expanded significantly. By the mid 1960s, however, as discussed in the last chapter, TSS was abandoned in favour of artificial regeneration methods. As well as pure tree plantations, Taungya farming was now promoted on a much larger scale (Lowe 1987). The species now planted were quick-growing exotics like teak (*Tectonis grandis*) and gmelina (*Gmelina arborea*) rather than indigenous hard woods, and maintenance of seedlings became the responsibility of the farmers themselves, who were paid for every surviving two-year-old seedling.[8] In this way Taungya farming was expanded rapidly and made an important contribution to feeding towns during the civil war (ibid.: 153). Overall, it seems to have worked quite well in this way until the mid 1970s (Darling 1995; Forestry Department 2002a).[9]

Narratives of the Decline of Taungya

Today it is widely accepted in Edo State that Taungya farming has, as one forestry officer in Benin put it, 'failed woefully'. The Forestry Department itself had two main explanations for this. Firstly, drastic cutbacks in government spending on forestry made the operation of the Taungya system logistically increasingly difficult. As outlined in previous chapters, there were two waves of severe retrenchment of staff, first in 1984 and then again in 1997, reducing the numbers of staff considerably. This is the case in Ovia South-West LGA, where 'there are no uniformed staff at present assigned to silvicultural duties'. Lack of vehicles and petrol has made seedling distribution impossible; there is only one serviceable Land Rover available throughout the state (Forestry Department 2002a). In fact, seedlings themselves are no longer available: in the 1990s tree nurseries have had to be abandoned as all members of nursery staff were made redundant.

In addition, forest officers generally blame farmers who, they say, often blatantly disregard the law. Foresters claim that even when they were still able to supply seedlings, which indeed was the case long after the mid 1970s, many farmers did not plant them, wilfully neglected the seedlings, or even 'planted them upside down', in order to deceive the forest guards. The Chief Conservator in Iguobazuwa also claimed that whereas in the past, farmers used to obey forestry rules and were 'very fearful, very obedient', this has now changed because like loggers and cocoa farmers, Taungya farmers now often have the support of politicians:

> Many farmers have a brother, son or friend in the House of Assembly in Benin, or as a Councillor or Chairman in the Local Government ... Because of that they look down upon the average civil servant who is supervising the Taungya system. They no longer take instructions seriously ... they defy the rules. And if you make any attempt to penalise them, you might lose your job.[10]

So today, the department claims, there is no way of controlling the farmers who want to farm the same piece of land again and are not interested in planting trees. In addition, there is also widespread illegal farming inside the reserves, which, like illegal cocoa farming and logging, is impossible for foresters with limited staff to detect:

> The farmers are wise now. Before, they do wait for the Forestry Department to give them farmland to farm. But today before you know it they go in there, take a really vast area of land for themselves, and start farming ... You can't discover them.[11]

Farmers themselves naturally have a somewhat different perspective. They emphasise the 'laxity', corruption and greed of forestry staff and regard higher-up officers as lazy and out of touch, never leaving their offices and unaware of what is happening on the ground. All readily admit that they do not plant trees any more, but the main reason given for this is that the Forestry Department does not supply them with seedlings. A group of farmers in Udo argued that they would plant trees if they could, but that the rotation system enforced by Forestry, of having to change every three years, was not favourable to planting. Several also said that they had only stopped planting quite recently: In Igueze, people planted teak until the early 1990s, whilst one farmer in Udo told me in 2002 that he and others still planted in the early 2000s: 'Only this year we stopped, because illegal people always bribe the Forestry people and cut them, so what's the point'. Furthermore, farmers say that most 'illegal' extension of Taungya farm land is undertaken with the unofficial approval of forest guards. To quote one Udo chief:

> Formerly, they (Forestry) were sharing it out 1, 2, or ½ acres per person. But now the forestry people they want money. They are just giving out the land, and wasting the forest. Forestry is very bad. The Head Office, the Director of Forestry, he will not know. It is the forest workers, such as the Chief Conservator, who are giving out the land ... At the end of the year, I will write as head of my group, and apply for 30 acres. This will be officially approved. But then, unofficially, I will ask and receive 20 more. So, 50 [acres] by 50 [acres], the forest disappears.[12]

These two accounts highlight quite different aspects of the breakdown of Taungya farming as a method of forest regeneration. Foresters do not mention their possible own role but instead overemphasise that of farmer's disobedience – many farmers did continue planting until quite recently. Farmers, in turn, are quiet about the fact that it suited them quite well not to plant trees anymore. Both accounts, however, support the general understanding of the decline of forestry in Nigeria. With dwindling government interest and spending making their jobs increasingly difficult and frustrating, the morale of forestry staff declined, whilst local power constellations further curtailed their capacity to implement policies properly. Most farmers in turn also lost motivation for carrying out their tree-planting duties, witnessing and taking advantage of the weakening of forestry staff. So over time both parties colluded to transform Taungya into quite a different, mutually beneficial arrangement. Now farmers have continuous access to good farmland, whilst Forestry staff have a lucrative source of income by being able to allocate the same piece of land repeatedly, and more of it than they are supposed to. These circumstances explain why the state government's two attempts in 1973 and 1984 to suspend Taungya farming altogether failed, as it 'succumbed to the pressure of farmers'(Forestry Department 2002a: 10).[13]

As a result, national and international observers alike find that Taungya farming has 'degenerated into ... a peasant shifting cultivation system which could eventually liquidate the forest reserves'(Oates 1995: 117) and has become 'a means of destroying the forest' (Forestry Department 2002a; Lowe 2000). Here, as with reservation and logging regulation, one sees environmental and political crisis narratives interlink and reinforce each other: administrative breakdown causes environmental destruction, whilst observed environmental destruction confirms a general image of lack of control and deliberate abuse by the Forestry Department. These views are again shared by local and international observers, even if there are clear discrepancies between foresters and farmers who blame each other for the system's failure. Because something went wrong, someone needs to be blamed. But what if one ceases to think of this breakdown as something wrong? For the tale of declining discipline is only part of the story. Underlying these developments was a continuous rise in farming population and demand for farmland which, I suggest, would have made a strict implementation of the Taungya system, even in its adapted form of the 1960s, quite unworkable by now. With this in mind, the next section shows that Taungya in its current form can be portrayed not as a failure but actually as quite a success, presenting a fairly realistic, socially accepted system of land allocation. In this respect, changes in the Taungya system, even more than those discussed in previous chapters, lend support to Scott's (1969) suggestion that corruption may also be understood as informal policy adaptation in a context where formal changes are slow or absent. Moreover, viewed from a wider perspective, the environmental impacts of the changes in Taungya farming have also been quite positive.

Alternative Perspectives on Taungya Farming Today

Although trees are no longer planted and often more land is allocated and farmed than regulations prescribe, in all other respects official Taungya procedures are still followed quite meticulously, displaying the same continuity in practices as contemporary logging regulations. Farmers of good social standing are still appointed as 'head farmers', who are in charge of a group of thirty or forty or so farmers.[14] Every January interested parties apply to head farmers for the amount of land they want, ranging mostly from one acre to three. Head farmers take the applications to the Forestry Department and receive a collective permit. When Taungya farms are allocated in February and March and divided up between individuals – not without disputes and arguments – all the plots are carefully measured out in chains, following a right-angled grid system. Like logging regulations that still more or less subscribe to rules set out in the early twentieth century, these are still the same procedures as before, following the original Taungya layout. It presents another visible legacy of the rationalised land management that scientific forestry endorsed, in sharp contrast with local, more 'organic' (Escobar 1999) land-use patterns under which farms of different shapes and sizes were scattered through the landscape.

Interactions with Forestry staff have, however, become more socialised than the original Taungya system perhaps envisioned. As one head farmer, showing me all the permits he had kept over the last six years, described it:

> You meet the Conservator of Forests once a year. When they come, they eat: fufu [pounded yam], hot [gin or other spirits], and so on. They bring the permit. In November, they send the assessment paper. On 20 December,

they say call your group, that they need some donation from them for Christmas, as a gift, one or two yams. By January, we pay ₦600. What they do with [it] does not concern us. Whether they chop, we don't mind.[15]

Kola nut too is offered to visiting forestry staff. Yam, hot and kola nut are traditionally signs of hospitality and respect amongst Bini speakers, used to signal goodwill and understanding between hosts and visitors (Ezele 2002a). Despite the fact that farmers speak negatively of Forestry staff (and vice versa), these rituals indicate a tacit understanding of each other's needs and help to facilitate and solidify their dealings; they build up patron–client like relations or 'friendships', similar to the relations and gift exchanges between cocoa farmers and local communities or between loggers and foresters.

In this socially embedded way, Taungya has played a significant role in the relatively smooth absorption of the large influx of 'strangers' and returnees into the Okomu area in recent years. Oates (1995) has argued that it was the availability of Taungya farmland itself that attracted people to the area, but this has been only one amongst several factors. As described in Chapter 3, the two foreign-owned plantations, OOPC and ORREL, employ a large number of workers from other parts of southern Nigeria. Virtually all current rubber and oil palm plantation workers, including those living inside the plantations, supplement their wages by farming. The plantations in turn also attracted more traders and business people to Udo, who sought to take advantage of the good road and general opportunities brought by Udo's development but who also often have a small farm.

These regionally specific factors coincided with rising food prices, caused by the country's worsening economy and Structural Adjustment Programmes implemented in the 1990s. High food prices have not only made food crop agriculture a profitable enterprise again, for smaller scale farmers as well as those setting up plantations, they also forced poorer urban dwellers to turn to farming, 'for survival'.[16] These included many retired or dismissed civil servants who were not receiving their pensions – in fact, a noticeable majority of people I encountered in villages and fields were quite old (see also Berry 1993). Because of the overall economic crisis, former traders and craftsmen in cities, and people who used to work for logging companies or on plantations, have also moved back to the villages to farm. Other Taungya farmers still live in Benin City, but have been given land in Okomu or Iguobazuwa Reserve through intermediate contractors, which they farm one or two days a week (see also Darling 1995).

All of this would have been more difficult and conflict-ridden without today's Taungya. Just as most communities have protected their land against larger-scale plantations, so many are reluctant to give land to strangers for small-scale farming. In Udo, Oliha and one or two other quarters have invited strangers to settle and farm, but several others, including Ogbe and Ihogbe quarters, only allow indigenous people to farm on their land. Through Taungya, reserve land is comparatively more accessible to outside farmers, just as it is easier to obtain for larger-scale plantation projects (see also Enabor, Okojie and Verinumbe 1982; Lowe 2003).

Of course, on Taungya too, indigenes may exert considerable control during land allocation: newly arrived strangers without any claims or ties to a village and with limited resources frequently end up with the worst pieces of land – only as they gain a stronger position within a community can they hope to get better allocations. Moreover, some local leaders make a considerable amount of money by providing access to Taungya land for strangers. OOPC workers, for example, can apply for Taungya land only through an Udo chief and have to pay a much higher price, up to ₦3,000–4,000 per acre. Nevertheless, Taungya enables strangers access to land they would otherwise not have.

Taungya has also made farmland more accessible to women. Traditionally, women farmed only on their husband's, father's or brother's land and did not own land themselves. Today, this is no longer the case as some farms on community land are also referred to as belonging to women; but Taungya land is more easily available than community land to a wider range of women, including unmarried, separated or widowed women, indigenes or strangers, who can apply for land in the same way as men. Here too, of course, there can be conflicts and problems for women. When new Taungya farms were allocated in Iguowan in January 2002, for example, I witnessed fierce disputes between the *odionwere* of Iguowan and a female farmer; she wanted to farm again in a plot that she had farmed five years before, but the *odionwere* insisted that he now had rights to this land. He was, in fact, resented by many Iguowan residents for frequently taking more land for himself than his allocated share. Overall, however, Taungya has enabled access to land to many, in a relatively smooth and conflict-free way.

At the same time, the environmental outcomes of Taungya in its present form have not been as destructive as normally claimed, especially when considering these both inside and outside reserves. In contrast to traditional farming, the creation of a new Taungya farm does involve the complete removal of all trees: another legacy of scientific forestry's desire for order and control. When they have just been cleared or harvested, Taungya farms appear barren and open, especially those that have already been farmed several times (Fig. 14). In this respect, they result in a more drastic reduction of biomass than any other forms of forest conversion. The actual area taken up by Taungya farms in Okomu Reserve, however, may be estimated to be approximately 45 square miles, whilst providing food and income for around 9,000 or more families.[17] This is significantly less than the area of the rubber and oil palm plantations (104 square miles). In Edo State as a whole the total area of reserve land

136 | *Things Fall Apart?*

Figure 14 A Taungya farm in Okomu Reserve. Photograph taken by the author, 22 January 2002.

under Taungya is, of course, considerably larger, even though it is again difficult to obtain precise figures. According to a 2002 Brief by the Forestry Department of Edo State, officially 10,000 hectares (38.6 square miles) are supposed to be opened up annually under Taungya, but given the various informal ways in which farmers obtain larger areas, and given my estimates for Okomu alone, it may well be twice as much, or more. Nevertheless, this area would be significantly larger, and expanding more rapidly, if Taungya farming still involved tree planting. Not planting trees allows farmers to return to farms after five years or less, much sooner than would be possible if teak or gmelina trees had to mature. This has greatly reduced the amount of new areas that need to be opened up each year. Some expansion still does take place; a farmer in Urheze, for example, pointed out to me in 2003 that the year after he would be farming in the high forest adjacent to his current plot. But on the whole, Taungya farmers in the north of Okomu Reserve were no longer clearing new land but returning to plots already farmed. In part, this was because little unconverted forest remained in this section of the reserve; such large areas had been taken up by the various plantations. But this only proves that, by the early 2000s, proper Taungya would have become quite unworkable: the one 'working', privately managed Taungya farm I saw, proudly shown to me by the Chief Conservator in Iguobazuwa Reserve, appeared to be rapidly eating into the remaining forest, leaving in its wake a rather unimpressive looking teak plantation. If all Taungya farms had continued to operate in this way, virtually all remaining high forest would have been converted years ago. This advantage of Taungya without tree planting – that it allows farmers to return to previously farmed plots much sooner – was, in fact, implicitly recognised in the 1994 Forest Utilization and Development Edict, which sought to restrict Taungya farming to areas already farmed (Forestry Department 2002a: 11): another instance where policy changes on the ground eventually work their way into official law.

Taungya in its present form has had even more positive environmental outcomes outside reserves. By making reserve land relatively easily available, it diffuses pressure on community land. Some community land has become quite depleted through years of repeated farming and shortening fallow periods. Around Udo and elsewhere large stretches of land along roads are now dominated by spear grass (*Imperata spp.*), which is very difficult to clear for farming. The fact that not all community land looks like this is largely due to Taungya. Access to Taungya land has enabled farmers to optimise land use on their own terms, allowing individual landowners to rest their own land for considerably longer periods than they could otherwise. On community land bordering Okomu Reserve this has resulted in something like a 'reverse reserve' effect: whilst reserve land has been cleared of forest, community land is covered in trees. Often large parts of community land are covered in rubber of varying age and quality – some old plantations have grown wild and contain a large number of different trees and plants apart from rubber (Fig. 15). But proper secondary forest too can be found on free land. Iguowan enclave, for example, in which people stopped farming in the 1960s when Taungya land became available and the village relocated to the road, is now quite forested, with only one or two food-crop farms in it (although a part of it has now been given to Yoruba cocoa farmers). In Udo the inhabitants of some quarters have been able to rest large parts of their land for fifteen to twenty years. Along the path to Ogbe quarter farms, for example, one walks through thick bush, with hundreds of different trees, shrubs and grasses that have great medicinal and other value to the villagers (Fig. 16).[18] As the Chief Conservator pointed out with some resentment, farmers were farming inside reserves whilst preserving their own forests.[19] Overall, these effects are clearly visible on a 1999 satellite image of the area, which shows much thicker vegetation just outside the reserve border than on Taungya farmland inside it (Fig. 17).[20] Taungya has therefore actually contributed to forest regeneration, albeit outside the reserves and with species composition that caters to villagers' rather than foresters' needs (see also Guyer and Richards 1996). In the case of non-indigenes farming in Okomu Reserve whose own land is elsewhere, such effects were more difficult to assess directly, but may have been similar. Several stranger farmers in Iguowan, for example, told me they were resting their own land for seven to eight years or longer.

138 | *Things Fall Apart?*

Figure 15 An overgrown old rubber plantation on Igueze community land. Photograph taken by the author, 9 January 2003.

Figure 16 Path leading through Udo community land in Ogbe quarter. Photograph taken by the author, 14 December 2002.

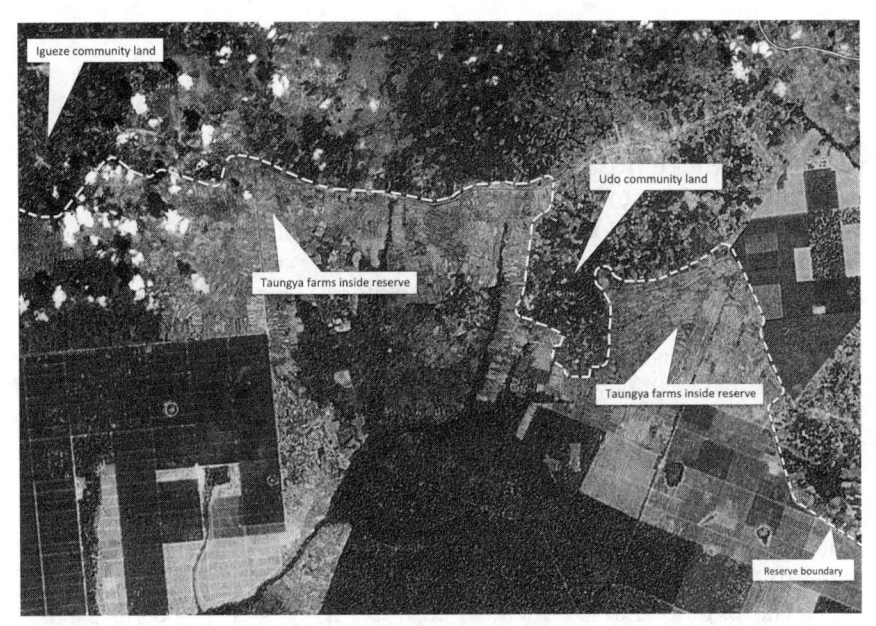

Figure 17 Landsat satellite image of the northern part of Okomu Reserve. Path: 190, Row: 056. Date of acquisition: 13 December 1999.
From Google Earth, drawing by Ricardo Leizaola, April 2013.

Overall, then, Taungya has not 'failed woefully'; on the contrary, its transformation may be seen as successful adaptation of a policy that would have been impossible to continue in its original format, with many positive social and environmental outcomes. However, Taungya farms too, whether allocated with or without the consent of forest workers, are in an insecure position vis-à-vis large-scale contractors dealing directly with the Director of Forestry, and are just as vulnerable to the expansion of large-scale plantations as cocoa and plantain farms. As already briefly discussed in Chapter 3, the village of Iguowan lost almost all of its farmland when Michelin expanded inside Okomu Reserve a few years ago (although it was allocated another compartment to farm in). Iguafole lost a large amount of Taungya land through the expansion of Mojo's farm, and Igueze farmers were similarly threatened when much of the land they were farming was given to Ogbomo. Most significantly, ORREL's expansion into Iguobazuwa Reserve in 2006 included a large area under Taungya farms, threatening the livelihoods of those farming there. It is therefore difficult to say how significant, ultimately, the benefits of Taungya farming will be in the long term, but in recent decades they have been substantial.

Conclusion: Things Fall Apart?

Together, the last three chapters have provided a more in-depth understanding of recent developments in Edo forestry than the common view that a once well-functioning system of forest management has collapsed due to lack of investment, management failures and widespread corruption. This re-examination has consisted of several strands of argument.

Firstly, it has shown that the Benin Forest Department never achieved its own environmental and developmental goals, namely the creation of a sustainable timber industry through the careful protection and scientific management of its forest estate. Forest policies did play an important role in the colonial reshaping of landscapes and economy, but in different ways than scientific forestry's technocratic vision intended. For one, they were far more political. Forest reservation played an integral part in colonial land politics, as an important means of land control for the *oba* against land privatisation through cash-crop production by urban elites. Between these emerging fault lines, rural communities were the biggest losers, as they saw community land rapidly taken away by both reserves and private plantations. The logging industry, meanwhile, also largely bolstered Benin's existing patrimonial political economy: chiefs accrued substantial benefits from timber royalties and made even more money as logging contractors, whilst the Native Administration's control over logging allocations became an important political tool for the *oba*. Taungya farming, in contrast, benefitted rural communities by providing much needed access to land.

Scientific forestry's environmental outcomes too fell short of foresters' hopes. Reservation did protect forests, but in doing so merely created separate spheres of forest and farmland that ultimately curtailed the regeneration of timber species both in and outside reserves. Timber regeneration was supposed to be achieved through the application of scientific methods of natural or artificial regeneration, but for logistic as well as ecological reasons, none of these were fully successful. An exception to this were early Taungya farms, on which planted and naturally germinating timber trees grew well – Taungya, of course, came closest to creating the conditions of shifting cultivation under which timber species had grown in Benin forests throughout history. Yet even with Taungya, foresters stopped short of fully integrating the insights many of them gained over the years they spent in Benin forests; in outlook and principle scientific forestry remained fully committed to government control over the forest estate and the application of scientific models and templates largely developed elsewhere. These, ultimately, were not appropriate for Benin forests.

Secondly, this re-examination showed that scientific forestry's legacies played a significant role in Edo forestry's recent decline. Reservation itself had given the state government control over reserves, which ultimately facilitated the use of reserve land allocation as a source of political patronage, whilst many (if not all) rural communities were able to protect their land more successfully. In the timber industry, illegal logging, corruption and the strategic allocation of concessions to political supporters did not begin in recent decades but had a long history. But even more important than these continuities in land politics and policy practices are the broader environmental and developmental changes brought about by colonial resource management. Forest reservation coincided with the emergence of cash-crop production, and from the 1930s there were conflicts between forestry, farmers and cash-crop producers and between the Forest and the Agricultural Department. Dereservation already began in the 1930s, and pressure for the release of reserve land for cash-crop plantations grew steadily from the 1950s onwards. At the same time, the Forest Department had facilitated the creation of the logging industry, even if this was always shaped more by market and technical factors than foresters' activities. By the 1970s the then Bendel State had a large and flourishing timber industry, whilst years of logging had already depleted Benin forests of much of its original timber stock. Because of ongoing regeneration problems, working plans were already then beginning to be disregarded. The demand for Taungya land, meanwhile, had risen to such levels that larger areas of new reserve were opened each year. Even under optimal political and economic circumstances, therefore, it would have been increasingly difficult and problematic to protect all reserve land against agricultural expansion, to meet timber demands, to maintain existing working plans, and to continue Taungya on its proper lines.

This is not to say that the severe economic and political problems that beset Nigeria from the 1980s onwards had no impact on forestry. Lack of investment and political support, staff retrenchment, low morale and widespread corruption clearly have contributed to incursions into reserve land, illegal logging, the collapse of working plans and the abandonment of tree nurseries, government plantations and Taungya planting. Nevertheless, recent decades do not present an abrupt change from rational, well-functioning management to complete collapse; rather, they need to be viewed in the context of previous developments, as the last of several stages in a long and often problematic history. Just as independence from colonial rule in 1960 was not a major turning point in this history, so the routine narratives that associate forestry's collapse with the end of oil wealth in the late 1970s alone need to be reviewed.

Thirdly, contemporary practices themselves have been closely examined, providing more nuanced insights into the nature and manifestations of patronage and corruption in the Nigerian forestry sector than standard narratives. Reserve land and logging licenses have indeed become sources of patronage, especially when their allocation has come under the direct control of the State Governor; political allies and financial backers often have first choice of logging and land allocations. There are also numerous other semi-legal and illegal arrangements between forest officers and loggers and farmers. As part of these, forest guards and officers are routinely given bribes, gifts or 'dashes' of various sizes in order to tolerate illegal practices. All this is framed in terms of social connections and 'friendships'; it is ultimately social standing, albeit often created through strategic donations, which enables some loggers or planters to operate the system much more effectively than others. In general, official procedures in logging regulation and Taungya land allocation are embedded in social interactions involving the exchange of small gifts, such as calendars, between logging contractors and forest officers and the traditional offering of kola nut, gin and yam by farmers. These different aspects of forest management practices today are rooted, on the one hand, in local history: in Benin's long history of patrimonial resource control and patron–client relationships and in customary social interaction. On the other hand, they are rooted in scientific forestry itself, in particular in its approaches to resource control (licensing, stamping, etc) that have proven prone to abuse throughout the world, in Europe as well as the tropics. They evolved as a result of interaction between the two and as a result of the process of forest management's establishment in the Benin Division.

However, it is important to emphasise that not all current practices are determined by the same principles. Just as in the Benin Kingdom, the *oba*'s patrimonial rule was always counterbalanced by communities' relative autonomy in local government and land management, so today patrimonialism and patronage politics are not the only forms of logic behind forest management practices. For one, not all forest staff, loggers and farmers participate in the same way; some forest guards and officers try to maintain official practices and do not participate in corrupt dealings, whilst some local communities have taken measures to control illegal logging themselves. Moreover, today's divergences from official rules can also be interpreted quite differently, as 'bottom-up' policy alterations in response to changing social needs, in a situation where official reform is slow or absent. This argument can be made especially with regard to the expansion of small-scale farming inside reserves, namely cocoa, plantain and Taungya farms – the very developments most decried as 'illegal' and failures of management. In the context of rising food prices, decreasing opportunities in cities and therefore few alternative means of obtaining food, they reflect real needs for land and 'survival' farming. They have made important contributions to local livelihoods and rural development, and, in Okomu, have facilitated the peaceful absorption of a large number of strangers. In contrast, the larger plantations in Okomu, whilst providing employment and contributing to regional development, have not only played a central role in Udo's recent communal conflict, they have also led to the displacement of many small-scale farmers.

Crucially, the environmental outcomes of policy alterations by small-scale farmers have also been far less destructive than those of larger-scale plantations. Cocoa and plantain farms maintain substantial parts of the original forest cover and do not drastically reduce biomass or biodiversity. Taungya farms are cleared completely of tree cover, but Taungya farming inside reserves has enabled farmers to allow long-term fallow and regeneration of forest outside reserves, on community land. Highlighting these positive outcomes of recent deviations from official policy presents another important corrective to routine assumptions that in the past forests were well protected and that recent management failures have resulted in forest destruction; whilst past practices had more negative outcomes than is usually assumed, recent changes have not all been inevitably destructive.

On the ground, alterations of official policies – 'illegal' farming inside reserves, the continuation of Taungya without tree planting – have provided rural farmers with space to pursue their own land management and conservation strategies. Communities traditionally had almost complete control over their land and resource use, a control that was much diminished through logging regulations and in particular the creation of reserves. Of course, communities retained control over unreserved land, but its reduction in size forced them to alter traditional extensive farming patterns. This did not prevent all forms of traditional conservation – groves around shrines, for example, have been largely maintained. Communities have also developed new conservation strategies on their own land in recent years, such as the planting of trees on individual farms or in communally managed tree nurseries. But Taungya has provided more opportunities for developing these strategies, just as the availability of reserve land enabled local and in particular migrant farmers to develop the 'new method' of plantain farming, with so many environmental and economic benefits. It is in part in this creative interaction with forest regulations, then, that one finds conservation practices that have developed organically on the ground, something that is missed entirely in conventional accounts of the decline of forest management. It is also largely missed in new initiatives to create community conservation in Okomu, the subject of the next chapter. Yet it is important to understand and draw attention to these local conservation dynamics, all the more so because they are increasingly threatened and replaced by the expansion of larger plantations. Ultimately, the legal ambiguities of land control and flexibility in law application that have provided rural farmers with many opportunities in Okomu Reserve also present their greatest vulnerability: they have little protection against larger, more powerful interests.

Notes

1. There do not seem to have been any significant differences between Chena and Taungya; at Sapoba it continued to be referred to as Chena, but later it was only known as Taungya.

2. NAI, IB FOR DEP 1/760, 223 WP, 1697 D, Annual Report of the Benin Native Administration Forest Circle 1935, 10.
3. NAI, BD 27/ X, Annual Report of the Benin Native Administration Forest Circle 1943.
4. Ibid.
5. NAI, IB FOR DEP 1/760, 223 WP, 1697 D, Annual Report of the Benin Native Administration Forest Circle 1935.
6. NAI, BEN DIV 6, BD 27/XV, Annual Report on the Benin Division 1948.
7. Ibid., 9. See also BEN DIV 6, BD 27/XI, Annual Report on the Benin Division 1944; 27/XIV, Annual Report on the Benin Division 1947.
8. Interview with a retired forest ranger, 20 February 2002. See also Darling (1995: 18).
9. Interview with the Chief Conservator of Ovia South-West LGA, Iguobazuwa, 17 December 2002.
10. Ibid.
11. Ibid.
12. Interview with Chief *Oliha N'Udo*, Udo, 9 December 2002.
13. Members of the Udo community wrote a letter to the Ministry of Agriculture and Natural Resources in March 1985. In this letter the community protested against the cessation of the Taungya system, arguing that it was a vital means of livelihood to them. Shown to me by Chief *Oliha N'Udo*.
14. Originally, this term referred to silvicultural assistants who were trained to take charge of farming centres, but now the term refers to anyone who is put in charge by the community. In effect, some of these are actually retrenched forestry workers who have continued in their positions.
15. Interview with Sunday Osayi, Udo, 10 December 2002.
16. I heard many different such stories of farmers in Iguowan, who had all been working in Benin City, Lagos or other places before. These particular words, 'for survival', are those of the Chief Conservator in Iguobazuwa.
17. It must be stressed that these figures are very rough estimates only; because they are not accurate, I have been deliberately conservative (i.e., maximising the extent of Taungya land). The extent of Taungya farms in the north could be estimated relatively accurately through maps and oral research; they amounted to around seventeen square miles in the early 2000s. Maps for the west and south were not reliable at recording the actual extent of land allocated under the Taungya system (for plantain as well as food-crop farms), but as a rough, generous guess I suggest they take up another twenty-eight square miles or so (fourteen in the south, which I visited several times, and fourteen in the west, which I was never able to visit). The number of families supported through Taungya farming in Okomu Reserve is based on the fact that most families farm between one and two acres, but a few farm more; plantain farms in the south can be seven or eight acres. I based my calculations on a generous average of farm size of around three acres to account for this minority of larger farms, or 200 farms per square mile.
18. This includes young people. During a walk along the path to Ogbe quarter farms, Susanna, a young, fashionably dressed college girl studying computer science in Benin City, was astonishingly knowledgeable about the names and uses of many different plans.
19. Interview with the Chief Conservator of Forests, Ovia South-West LGA, Iguobazuwa, 14 January 2003.
20. I was originally given this picture by a manager at OOPC.

CHAPTER 5

Okomu National Park
A Postscript on Conservation

Introduction

International tropical forestry has undergone significant shifts in recent decades. Historically, its main framework was one of centralised resource management, focusing in particular on timber and fuelwood production. In mountainous and drier regions it also had an environmental or conservation rationale, which emphasised its importance for watershed protection or for combating soil erosion and desertification. But overall the resource framework prevailed, particularly in naturally forested and timber-producing areas such as southern Nigeria. In recent years, however, forestry's priorities have shifted in that the protection of biodiversity, of flora as well as fauna, has become an equally important agenda. On the whole, forestry has become more conservation oriented, even where timber production and the regulation of logging activities remain key tasks.

At the same time, informed by broader political changes as well as the growing recognition of the various shortcomings of centralised approaches, there has been a global move towards decentralisation, with more control over forests being given to local or regional rather than central authorities (Colfer, Dahal and Capistrano 2008; German, Karsenty and Tiani 2010; Ribot and Larson 2005). This shift has taken place not just in forestry departments but also in wildlife conservation, as the 'fortress conservation' (Brockington 2002) of national parks is being replaced or adapted by various community conservation programmes. These have taken a number of different forms, ranging from 'conservation-with-development', such as park outreach programmes implemented by existing national parks, which provide various kinds of compensation for local communities to ensure their cooperation, as well as 'community-based natural resource management', whereby local communities themselves are in control of wildlife management (Adams and Hulme 2001a, 2001b).

To its proponents, community conservation promises to make conservation more effective whilst simultaneously improving local livelihoods, but its various strategies, in particular conservation-with-development, have also received considerable criticism. Conservationists argue that too much of an emphasis on development can be counterproductive and endanger the very conservation aims of a project (Oates 1995), whilst those more concerned about local people's welfare see it as mere window dressing that allows the continuation of existing conservation approaches and denies local people meaningful control over natural resources; a 'shallow and perhaps deceitful façade designed to hide old-style preservation' (Adams and Hulme 2001b: 193).

In Nigeria, largely as a consequence of the economic and political difficulties of the last decades and the country's increasing isolation from global initiatives, forestry has been slower than in other African countries to adopt community forestry as an official policy. In the north, local government authorities officially work in partnership with state governments, but in the south, they are not involved (EC-FAO 2003). Overall, there has so far been little attempt to officially devolve management towards local authorities at a community level. Forestry's reorientation towards biodiversity and conservation too has been slower than elsewhere, but is gradually taking place now. In Edo State the Forestry Department remained under the Ministry of Agriculture and Natural Resources (MANR) until 2006, when it was put under the Ministry of Environment.

Nigeria has participated somewhat more in broader regional trends of wildlife conservation. Generally, wildlife conservation has a longer tradition in the colonial hunting grounds of East and Southern Africa, where large conservation areas and national parks were established from the late nineteenth century onwards (Anderson and Grove 1987a; MacKenzie 1989). After independence the creation of further National Parks formed an important development strategy in countries like Kenya and Tanzania (Adams and Hulme 2001a; Neumann 1998). In West Africa some game reserves were created in the colonial period – in the Benin Division, in fact, the Gilli-Gilli Game Reserve was the very first reserve, provisionally created in 1907 – but because both sporting colonists and large game were sparser on the ground,[1] they never obtained the iconic importance of their counterparts in East and Southern Africa. Consequently, West African national parks appeared much later: in Nigeria the first national park, Kainji Lake in Niger State, was established in 1979, and the National Parks Governing Board and five further national parks in 1991 (Marguba 2000).[2] At the same time, international donor agencies and local NGOs have been promoting various community conservation programmes, in particular in the category of conservation-with-development. The most important sites here were Korup National Park in southern Cameroon and, just across the border, Cross River National Park in southeastern Nigeria, where WWF-UK and other bodies set up a series of park outreach programmes to involve local communities and foster sustainable development (Ite 1998, 2000; Malleson 1999, 2002). A similar conservation-with-development programme, funded by the same donor agencies and involving some of the same conservationists, was set up in Okomu Reserve. Here, the Okomu Wildlife Sanctuary was created in 1985 and became Okomu National Park in 1999.

This final chapter looks at these recent initiatives and at the role that the sanctuary and the national park have played in the local ecology and political economy. It is a key premise of this book that, as such community-based conservation and forest management programmes are being promoted and adopted throughout Africa, it is all the more important to know the longer-term history of forest management. This knowledge helps not only to gain a more nuanced understanding of the various problems (real and perceived) existing forest management has run into by today, but also to assess the recent initiatives seeking to replace former approaches. Here, the knowledge of past and present forest relations at Okomu built up in previous chapters enables us to situate the sanctuary and national park in their wider historical and regional context. On this basis, the chapter begins with a critical analysis of the rhetoric and ideas that accompanied the conservation-with-development programmes adopted at the time of the wildlife sanctuary. Drawing on Neumann's (1997) critique of the 'primitiveness' of ideas underlying community conservation discourse, it shows that the settlements inside Okomu were portrayed as more 'traditional' than they really were. At the same time, existing local land management and conservation practices such as described in the previous chapters – the development of the comparatively sustainable 'new method' of plantain cultivation;[3] the protection, fostering and planting of timber trees on farmland; the long fallow periods and growth of secondary forest on community land enabled by the availability of Taungya farmland – were ignored; like elsewhere, the content of both conservation and development ideas were largely imported rather than locally generated. Consequently, the outcomes of conservation-with-development at Okomu have been neither as successful as its proponents had hoped nor as detrimental as its critics had feared. The national park that succeeded the wildlife sanctuary in 1999 has also been less significant in the local political economy than is often assumed in the literature on national parks. However, it is in part *because* of its smaller role that it has been relatively successful in terms of environmental protection and local acceptance. The chapter ends by briefly reflecting on the implications of the findings of this and previous chapters for future forest conservation policies.

The Okomu Wildlife Sanctuary

In 1982 the conservationist John Oates of Hunter College, New York, and P.A. Anadu from the University of Benin went on a two-month tour through southwestern Nigeria funded by the World Wildlife Fund (WWF), surveying its forests and wildlife populations. Their aim was to gather information that could form the basis of conservation proposals. John Oates was particularly interested in the rare white-throated guenon monkey, *Cercopithecus eryrogaster*, which is endemic to southwestern Nigeria. It was in Okomu Reserve that they spotted most white-throated monkeys. Since Okomu also contained a high diversity of interesting lowland rainforest mammals and birds, and since it was the 'largest and least disturbed remaining example of a typical southwest Nigerian rainforest ecosystem'(Oates and Anadu 1982: 1), now under threat from the recently established Okomu Oil Palm Company, it was here that Oates and Anadu proposed a wildlife sanctuary be created.[4] They recommended that this sanctuary should cover seventy-four compartments (190 square miles) in the centre of the reserve, bounded by Okomu River in the west and the Udo-Nikrowa road in the east. They submitted their proposals to the Bendel State Government, and, with a three-year delay and, at twenty-six square miles, significantly smaller than Oates and Anadu had hoped, the state government formally approved the creation of the Okomu Wildlife Sanctuary in 1985, in the heart of Okomu Reserve.[5] It was to be jointly managed by the Ministry of Agriculture and Natural Resources (MANR) and the Nigerian Conservation Foundation (NCF). The Nigerian branch of the A.G. Leventis Foundation, which ran a number of agricultural development and conservation programmes in Nigeria, was also involved in setting up the project. The main donor agencies were the WWF and the Overseas Development Agency (ODA) – ample donor funding no doubt played a significant role in persuading the state government to agree to such a project at a time of increasing financial difficulties.

The sanctuary's main purpose was to protect the white-throated monkey and its forest habitat, and Oates himself, as a motivating force, had little interest in community involvement. Much to his irritation, the NCF as well as the WWF and ODA had different ideas. They believed that the sanctuary's survival depended on 'integrated management' of the whole reserve and community development, in order to ensure local cooperation and goodwill and to compensate local farmers for their loss of access to land. The type of community development envisioned included a move towards more 'sustainable', intensive farming methods, and the creation of alternative sources of income.[6] In 1990 five ODA-funded British consultants, including a forest management expert, a soil specialist, an agriculturalist and a wildlife specialist, came out to Okomu to identify possible strategies for achieving such 'integrated management'.[7] In particular, they identified the need to control Taungya farming as crucial for achieving sustainable farming and successful conservation in Okomu.

This brief visit was followed by a more in-depth, ODA-funded socio-economic survey of Okomu Reserve, which was conducted in 1991 and 1992 by the sociologist Francisca Omorodion of the University of Benin in cooperation with Peter Lloyd, Professor of Social Anthropology at the University of Sussex. Adopting a similar approach to the programmes at Korup and Cross River, the survey was to be part of the preparation for improved forest management and conservation in Okomu Reserve, under what was now called the Bendel State Forest Project (Lloyd 1991). A preliminary visit to Okomu had already given Lloyd 'a vivid impression of the scale of forest destruction and degradation and of the pressures placed on the remaining high forest by farmers and hunters and others' (ibid.: 4). The aim of the survey was, therefore, to provide a descriptive baseline for the formulation of policies introducing new modes of farming and new income-generating activities (IGAs) to the area.

Francisca Omorodion first conducted a pilot study of four villages – Iguowan, Arakhuan, Mile 3 and Nikrowa – in December 1991. Her report is in many ways typical of the language and ideas of the 'community conservation' paradigm. She begins by stating that 'the forest is of great significance in all aspects of people's lives, providing foods, fuelwood, medicines, craft materials, building materials, etc. The forest also contains deities that are believed to culturally influence the productivity and well-being of the people' (Omorodion 1991). Here, Omorodion conjures up a well-known scenario of indigenous people whose material well-being and spiritual identity depend on the forest, a scenario that plays an important role in community conservation discourses. But this pictured 'primitive-ness', as Neumann (1997) has pointed out, is often misleading, in Okomu as much as elsewhere. Already in the early 1990s a large percentage of people living and farming in Okomu Reserve were from other parts of Nigeria: former timber workers from Eastern Nigeria remaining in the reserve to farm, people from other parts of Edo State engaged in Taungya farming and Yoruba cocoa farmers who were beginning to arrive at this time. These people depended on the availability of farmland and may well have used forest plants for medicine, but to them the forest had little spiritual meaning. Of course, to indigenous Edo and Ijaw people parts of the forest, such as Iguowan Lake or Okomu River, did contain deities (as discussed at greater length in Chapter 2), and many also used forest trees and plants for medicinal and other purposes, particularly those living in the direct vicinity of high forest, such as in Iguowan or Mile 3. However, indigenous people too did not live traditional, subsistence-based lives but participated in the larger regional economy. Udo residents, for example, now usually buy non-timber products traditionally collected from the forest, such as giant snail (a local delicacy) or the twigs for making 'outside' brooms, at the market, as this is more convenient than lengthy trips into the forest.

Omorodion goes on to say that due to the non-availability of farmland fallow periods have decreased from seven to three years, and finds that:

> [T]he majority of people are against the conservation of the forest, viewing it as responsible for the low soil fertility, non-availability of high forest for farming and inability to obtain animal protein through hunting of bush meat. The people are aware of the prohibitions arising from the 'reserve' status of the forest but lament the failure by government and other NGOs interested in the area to recognise the significance of the forest for their livelihood and therefore, the need and importance of providing alternative IGA and social amenities that will ameliorate their sufferings. It is paramount that unless the government and NCF gain the support of the villagers, there is a risk that the reserve status of the forest will continue to create conflicts, discord and lack of co-operation between those focussing on conservation and the local people in the area. (Ibid.)

Again, her analysis largely conforms with others elsewhere, namely that conservation impacts local livelihoods negatively, and that, for conservation to be locally accepted, alternative IGAs are necessary in order to 'ameliorate their sufferings'. Indeed, she recommends that the NCF promotes the establishment of fishing ponds and breeding centres, replanting through the Taungya method and conservation education – all typical of community conservation programmes. However, this analysis of the problem and the suggested solutions are again not quite correct. For a start, she appears to conflate the larger, long-existing Okomu Reserve with the newly created conservation area of the wildlife sanctuary. This creates the impression that people are denied access to the whole reserve, which is not strictly true: in the larger reserve they could gain access to Taungya farmland, hunt and collect firewood and medicine. At the same time, the report suggests that local farmers present the main threat to conservation and ignores the presence of OOPC, which, after all, presented a much bigger threat to both farmers and conservation than they did to each other. Moreover, the alternative IGAs suggested do not present a viable alternative to farming as a source of income; whilst beekeeping, fish ponds and poultry can provide valuable income and nutrition supplements, they cannot fully substitute that derived from farming and would not therefore significantly curtail it. Finally, the survey is also typical of most community conservation initiatives in terms of conceptualising local communities simply as recipients; both the conservation and development elements in the programme were conceived by outside experts. Even if local communities were nominally consulted, the overall approach is still largely top down, just like centralised forest management before it. But as I have tried to show throughout this book, rather than passive recipients, local people are agents who for many years have already been actively reshaping and managing their environments. Local land-management practices on community land, however, do not feature in the report; just like that of scientific foresters before, the conservation-with-development lens focuses on the reserve alone.

In the final report that she and Lloyd wrote in December 1992, after Omorodion had conducted a wider survey, some of these earlier limitations are recognised (Lloyd and Omorodion 1992). They acknowledge the mobility and urban linkages of most village residents and point out that most Taungya farming takes place directly adjacent to the enclaves. It therefore does not directly threaten the sanctuary. Moreover, they recognise the much larger role of logging activities and plantations in forest destruction, and therefore change their focus considerably. They emphasise the need for development to 'involve the full participation of the people living in the reserve', but now speak of the need to bring together all interest groups that hold 'stakes' in Okomu Reserve, including plantation and logging company managers, and to reach a consensus between them. Given the complexity of the situation, their recommendations amount merely to suggesting that, again, technical experts – agriculturalists, foresters, zoologists, social scientists etc. – with knowledge of local practice, research and problems should be consulted to establish the most appropriate modes of development to achieve the goals established.

In the meantime, however, the NCF had already gone ahead with its own community development programme. In addition to the management of the wildlife sanctuary, NCF staff had enthusiastically begun to see themselves as responsible for bringing about sustainable land-use practices not only in the envisioned buffer zone around the sanctuary but in the reserve as a whole. This was the Okomu Forest Project (OFP), which was partially funded by the Ford Foundation. An NCF Support Zone Officer was appointed, and various different programmes were embarked upon, with the overall aim of encouraging intensification of community economic activities through education and the establishment of a revolving credit fund. Improved IGAs were introduced, such as, for example, a poultry project at Iguowan. For this, 6,000 day-old birds were bought to be resold to farmers, with a lecture on poultry management delivered at a special workshop that covered breeds, vaccination, medication and feeding. A cassava mill was also given to Iguowan as a loan, to be run by women. Demonstrations of farming with improved high yielding varieties of maize and cassava were held and a beekeeping project introduced. In Nikrowa similar demonstrations were held with HYV maize, and a piggery was established. A programme of environmental education was also launched, which involved several meetings with men and women from different communities in and around Okomu Reserve (Egbhuche et al. 1995).

However, the OFP soon ran into difficulties. Peter Lloyd had in fact in all of his reports already expressed some doubts about the ability of the

NCF to implement such an ambitious programme of rural development, given their relative lack of experience in agriculture and rural affairs. These doubts seem to have been justified. Nick Ashton-Jones, West and Central Africa representative of Pro-Natura International reported to the NCF after a visit to Okomu in March 1993 that 'shortage of funds, an ill-defined policy and confused management ... resulted in the poor performance of the project, low morale amongst staff, loss of confidence in the project from potential supporters and an increase of timber poaching from the Sanctuary' (Ashton-Jones 1993). The Ford Foundation too was dissatisfied; most of its funds of $60,000 had been disbursed, but hardly any loans had been recovered and little had been achieved. Worse, they feared that the project may have been counterproductive in that expectations were raised only to be dashed again (ibid.).

During my fieldwork in 2001–2003, I saw only few traces of the beekeeping or poultry projects. At Iguowan some people made honey and a small number of chickens were described as 'agric', which may well have been descendents of those introduced by the NCF. However, neither honey making nor 'agric' chickens were of economic importance or part of larger local projects. Only the cassava mill, still up and running, was important to people's lives; everyone said that it was 'very good' of the sanctuary to have provided it. However, this one cassava mill at Iguowan neither significantly increased people's incomes nor did it lead to the kind of large-scale immigration of farmers that Oates, the fiercest critic of these local development programmes, had predicted (Oates 1995). The overall impact of the Okomu Forest Project, therefore, has been less than its advocates had hoped or its critics had feared. In Ijaw villages south of the sanctuary the project may have raised expectations that it did not fulfil, causing anger and resentment against the sanctuary that people still felt in the early 2000s.[8] In the larger context of Okomu Reserve, however, the wildlife sanctuary and its development programme were simply not so central to people's lives that its failure made a huge difference – they continued with their normal farming activities, which had always taken place far away from the centre of the reserve where the sanctuary was created.

Eventually, like Lloyd and Omorodion in their final report, the project leaders became aware of the fact that their community development efforts were not only ineffective but also misdirected with regard to achieving forest conservation. From 1993 onwards attention began to shift to the threatening further expansion of the plantations, in particular to that of the Osse River Rubber Estates (ORREL), recently taken over by Michelin. As described in Chapter 3, Michelin until then only had land outside the reserve, but in 1993 its management was in the process of acquiring a further 2,000 hectares inside the reserve. Both Iyayi and OOPC were expanding too, but the expansion of Michelin was particularly worrying because their newly acquired land was immediately adjacent to the sanctuary and also surrounded all of Iguowan, the settlement where the NCF was based. A cost–benefit analysis was prepared on behalf of the NCF, during which alternative forms of land use, i.e. protected natural high forest and rubber plantations, were compared. It came to the conclusion that, mainly due to the precarious nature of the international rubber market, natural high forest areas had a higher economic return than rubber plantations, and that rubber plantations should be planted not in high forest, but outside reserves (Coates et al. 1993). This cost–benefit analysis however did not prevent the allocation of the 2,000 hectares of reserve land to Michelin, so instead the NCF now started a campaign to persuade Michelin to leave the two compartments directly adjacent to the sanctuary untouched as a buffer zone, and also to leave some land and high forest around Iguowan. This campaign involved an ILO lawyer and the archaeologist Patrick Darling, who even managed to attract the attention of the *Financial Times* (Adams 1995). The campaign was successful: although Michelin did take over compartments 20 and 21, previously farmed by Iguowan residents, it left Iguowan Lake, the area east of the village identified as sacred, as well as a little strip of forest on the western side of the village. It also left compartments 28 and 29 as a buffer zone for the sanctuary. Although the buffer zone theoretically also prevented local communities from farming in these compartments, its primary role here was to act against the expansion of the plantations. Because it was also established on already reserved land, it did not present an incursion of state control into community land, such as described elsewhere (Neumann 1997).

In general, the NCF now recognised the need to cooperate with all the 'stakeholders' of Okomu Reserve, including the plantations as well as the communities of the area, in order to achieve its integrated management. Meetings were regularly being held with representatives from Udo, communities inside the reserve, OOPC and ORREL, and stakeholder participation formed the cornerstone of the next report on Okomu Reserve, produced in 1995 by Patrick Darling (1995). This 'master plan' draws a more realistic picture of the various activities in the reserve, including legal and illegal logging and the mechanisms by which land was allocated to rich individuals. However, by focusing on the reserve alone it again does not take land management practices on community land into consideration.

The plan recommended buffer zone management, intensification of agriculture, environmental education and development of tourism. Whilst tourism remained limited and agricultural practices changed little, environmental education workshops were held quite regularly, according to Udo and village residents. The two plantations also agreed to the creation of buffer zones: OOPC agreed to keep buffer zones of high forest along the border with the park and along the rivers going through the plantation, whilst ORREL, in addition to not planting the two compartments adjacent to the national park, pledged to preserve forest around rivers and lakes within its plantation. OOPC also contributed to road maintenance and provided a few vehicles to the sanctuary.

Indeed, both plantation managements, keen to improve their local and international reputation, now sought to portray themselves as environmentally aware and supportive of conservation. Managers frequently pointed out to me the ecological advantages of rubber and oil palm plantations, such as soil protection and increase in biomass. They also drew my attention to the protected areas inside their plantations and to their support of first the wildlife sanctuary and then the national park. This they contrasted with the uncontrolled expansion of farmland by local farmers and the destruction of forest this brings with it, pointing out the conversion of forest into farmland along the road towards OOPC that had taken place 'only in the last few years'. If they, the environmentally conscious plantation managers, were not here, the whole place would soon look like this, I was told. Managers did not mention, however, that the people farming on this particular stretch were in fact their own workers. In general, these declarations of environmental concern cannot be taken at face value: as we will discuss below, some of their conservation commitments were rather short-lived and remained secondary to their economic interest. Nevertheless, it was in the interests of the plantations to at least give the appearance of supporting conservation initiatives, and the kind of wildlife conservation undertaken by the sanctuary would have been a familiar and respectable concept to them. For these reasons, large-scale plantations can potentially work quite well with conservation projects, as Butler and Laurance (2008) have suggested.

The master plan also recommended the establishment of a national park, something the NCF had been long been campaigning for. With continuous lobbying, this finally took place in May 1999, in a handing-over ceremony held at the park's administrative headquarters at Arakhuan.[9]

Okomu National Park

The transformation of the wildlife sanctuary into a national park has brought with it several changes. The NCF handed over control to the National Parks Board, who now employed their own officer and ranger staff, although several NCF forest guards were kept on.[10] Staff size increased considerably: in the early 2000s there were several specialist officers for areas such tourism, research and environmental education, an engineer, two accountants and about eight clerical staff as well as forty rangers, all under an Officer in Charge. The national park also has the right to arrest and prosecute poachers and illegal loggers, which the NCF sanctuary did not have. This has considerably increased their ability to protect the forest and wildlife of the national park, especially since all national parks had received paramilitary status with the 1999 National Parks Decree. In 2000 seventy-five national park officers and rangers drawn from all eight national parks received three months of training in paramilitary techniques, tactics and survival skills at Yankari National Park (Hamman and Stopfords 2001).[11] Afterwards those Okomu officers and rangers who had attended the training trained the rest at Okomu. All in all, with a larger, trained and armed cadre of rangers and officers, protection at Okomu is now potentially more effective than it was under the NCF. Community development projects and stakeholder meetings, however, have been reduced considerably. In effect, Okomu has shifted from conservation-with-development to 'fortress conservation', something of a reverse trajectory to most conservation projects in Africa.

However, this has not led to a marked increase in local antagonism towards the park. There have been, it is true, some instances of protest. One particular cause of conflict was the park's planned extension by fifteen compartments (3,900 hectares) on its southern border. The extension was officially approved and marked on paper by 2001, but its actual demarcation on the ground has not yet fully taken place, due to resistance from local communities farming in this area. Farmers north as well as south of the park also frequently complain that the park's wildlife – especially monkeys, but also sometimes elephants – eats and destroys plantains and other crops on their farms. But the majority of local people, especially in the north of Okomu Reserve, claim to support the national park. In Iguowan, where many rangers and a few officers are stationed, relations between the community – already a mixture of indigenous people and people from other parts of Nigeria – and national park staff are relaxed and friendly, with much mingling and social interaction.

There are several reasons for this overall acceptance of the national park. At Iguowan a number of indigenes of the village are themselves national park employees, and here the park also provides the fuel for the generator donated by Michelin in 2001 ' – interestingly, the park receives more credit from Iguowan residents for providing the fuel than Michelin for the generator. In Udo and other surrounding communities there are no tangible benefits of the park's presence, but people still speak positively of it. I was often told that the national park was good because it would benefit everybody, even 'children yet unborn'; that because of the park, 'my pikhin [child] will know elephants'. These particular formulations clearly reflect the NCFs environmental education programmes, which may have resulted in some genuine appreciation of the 'animal kingdom', as the national park was frequently referred to. Such local discourses perhaps do not amount to the emergence of genuine 'environmentality' such as Agrawal (2005) describes in the Indian state of Kumaon, but, at the very least, they indicate that people know what kind of answers visiting researchers might like to hear.

However, there are also other reasons for people's acceptance of the park. For one, despite the national park's increased patrolling, capture and prosecution capacities, these are not necessarily always fully employed. A few rangers are always on patrol, and the park's log book shows that illegal loggers and poachers are caught quite frequently – about twenty times a year. Park management and most rangers I spoke to were adamant that illegal logging, which had been a problem during the time of the sanctuary, was no longer taking place, and indeed nothing was ever obviously visible along the main road or from any of the paths I was taken on by guides. However, rangers are also known to prefer 'patrolling the streets of Udo', as the Tourism Officer once put it, to patrolling the forest itself, and also to be lenient on poachers or hunters who were able to pay them. It is therefore possible that there is still some logging inside the park, even though this is vigorously denied by its staff. The expatriate managers of ORREL and OOPC I spoke to were convinced there was ongoing logging, pointing to the large number of timber lorries exiting the forest every day. These could equally have come from south of the park, but reports about illegal logging in the park continue to emerge.[12] Farming inside the national park, although isolated and small scale, is also effectively tolerated: those farming in the attempted extension in the south were not yet expelled in 2006, whilst Yoruba cocoa farmers, forced to abandon their farms elsewhere by the expansion of the OOPC and Mojo's farm, have started planting inside the national park. The park's guards know this, but do not seriously prosecute the farmers, preferring instead to collect small bribes when they come across the farmers.[13] On the whole, therefore, the walls of the fortress are not quite as impenetrable as its paramilitary status suggested.

The second reason why relations are relatively peaceful, at least north of the reserve, is that everybody knows that the national park is short of money and under-resourced. This is partly because of the limited government funding it receives, and partly because it generates virtually no income itself, despite ongoing attempts to develop eco-tourism here. In order to attract and accommodate larger numbers of tourists, the guest-houses at Arakhuan and at Mile 3 were repeatedly renovated by, in turn, the Bendel State Government and the Leventis Foundation: in 2006 an entire new building complex, with conference facilities and a swimming pool, was constructed at great expense. Maintenance of these buildings in humid forest conditions and with little staff is, however, difficult; previous buildingsall decayed quickly. Road maintenance too, is a problem, and in the rainy season the road often becomes entirely impassable. At any rate, due to Nigeria's underdeveloped tourism industry the number of visitors remains small. Okomu visitors rarely outnumber a hundred a year and consist mostly of Nigerian school classes, expatriate residents and a few bird watchers or others with specific interests in Okomu's wildlife. Overall the park is not very visible, has little money and does not present any obvious sources of income to local people.

In contrast, OOPC and ORREL play a significant role in the regional economy. Through contracts, gifts, jobs and at times extortion, they present lucrative sources of income and therefore receive far more attention than the national park. OOPC and ORREL have never-ending streams of visitors, their managers are frequently kidnapped and their vehicles stopped on the road, but the park is largely left alone.[14] With its headquarters at Arakhuan, in the middle of the forest, this was not surprising, but even after its headquarters moved to Udo, in 2005, few people dropped by there. The negative impacts of the plantations, too, are felt more keenly by local inhabitants. As described previously, it was the expansion of Michelin that threatened the existence of Iguowan and that of OOPC that forced cocoa and plantain farmers to leave, whilst several communities' Taungya land has been taken over by larger developers.

In the presence of the more important entities of the large plantations, therefore, the national park has not been a great engine of local development, but neither has it presented a threat or source of conflict. What, then, about its primary aim, conservation? Despite vigorous denial by park management, it is possible that some logging has taken place inside the park. Wildlife protection too may not have been as rigorously implemented as the national park regulations require. However, it is difficult to ascertain the scale and impact of hunting inside the park, considering that many animals move freely in and out of the park and can be shot or caught in traps outside its borders. On the whole, a large range of animals live at Okomu, including the white-throated monkey and other monkeys, forest elephants, buffalos, bush pigs, duikers and civet cats as well as a great variety of birds, butterflies and insects. Because no official count has ever been attempted, their overall numbers are not known, however. Animals are notoriously hard to spot in the forest; Okomu's elephants, for example, are rarely if ever seen, even though footprints and droppings bear testimony to their presence.

Conservationists involved in the Okomu project were concerned that the area was too small to offer sufficient protection and room for movement for the animals and birds inside. This was the main reason for the park's expansion. In 2002 The Edo State Chapter of the NCF also secured funding from the French Embassy for creating an additional protected area in the wetlands in the very south of the reserve. It was hoped that a corridor could be created between the park and these wetlands – a strategy that has become increasingly common in response to the shrinking amount of land available for conservation (Beier and Noss 1998). The initiative did not, however, result in any tangible outcomes.

It remains to be seen whether, in the long run, the national park in its current size will offer sufficient protection to its wildlife. Nevertheless, even with some logging and hunting of wildlife, it still plays a significant role in preventing forest conversion at the heart of the forest reserve, especially with regard to the further expansion of the large-scale plantations. As we have seen in Chapter 3, since 2003 rising rubber prices have induced both plantations to rapidly increase their planted areas. OOPC started clearing and planting the land it already owned, displacing the cocoa and plantain farmers who had been farming there, and in 2006 Michelin sought to clear the compartments it had previously agreed to leave forested as a buffer zone. A concerted campaign by the national park, the Edo State Forestry Department and the NCF prevented this – leading Michelin to acquire land in Iguobazuwa Reserve instead – but the episode clearly reveals the limits of the plantation's conservation mindedness. In this context the national park's unequivocal legal status as a protected area presents the only effective measure to prevent the large-scale plantations from expanding into the heart of the reserve. Smaller plantation developers, such as Mojo, and cocoa and plantain farmers are also less likely to attempt to farm inside the national park, even though there have been isolated incidents of cocoa farmers expanding their farms across the park's border. All in all, therefore, the park plays an important conservation role in Okomu Reserve.

Conclusion: Implications for Future Policy

The conservation-with-development rhetoric and programmes at Okomu followed well-trodden paths, including the evocation of familiar images of indigenous farmers' forest dependency and the introduction of alternative income-generating activities such as beekeeping and poultry. Discussing these at the end of this long term and holistic study of forest relations in the Okomu region has brought out their limited appreciation of the social, economic and political realities of the area. Its inhabitants lead far more connected, economically engaged and modern lives than the 'primitiveness' discourse allows for, and it is not surprising that attempts at community development through alternative IGAs largely failed. At Okomu conservation-with-development has not resulted in the kind of idyllic, sustainable rural development its advocates hoped for, but neither has it increased environmental destruction through the expansion of farming, as its critics feared.

The long term and regional context built up in previous chapters furthermore shows that the national park too plays a less significant role in the region than the literature generally tends to attribute to such conservation projects. Whilst the park has paramilitary status and the capacity to prosecute poachers and illegal loggers, in practice, it is often lenient. Moreover, it is completely dwarfed by the two large expatriate plantations, which offer far more lucrative sources of income. However, the park's relative economic and political unimportance has meant that it has been largely accepted locally, and that it has actually been able to fulfil its conservation role in the heart of Okomu Reserve, preventing the further conversion of forest by OOPC and ORREL. In the context of rapid expansion of large expatriate plantation projects, therefore, strong government protection of some parts of forest – the very principle that I critiqued with regard to large-scale colonial forest reservation – may after all be the most effective measure. The perhaps somewhat contradictory nature of these conclusions highlights the importance of finding locally appropriate conservation measures that are historically informed and take account of specific, and changing, local circumstances. There are no universal recipes; what works at Okomu now may not work here in the future, or anywhere else.

This, then, is one of the implications of this study for possible future directions in forest management. Scientific forestry consisted of the universal application of management templates that in practice encountered numerous difficulties, even when they were locally adapted. Yet many of its key tenets live on – still today, for example, official Nigerian forest documents state as a goal the reservation of 25 per cent of land, a formula now almost a century old that has become largely meaningless. Even in those parts of Africa where forest management has become decentralised and where community conservation has been promoted far more vigorously than in Nigeria, there has not been a complete departure from old forestry principles, which in substance if not implementation have often remained unchanged. Moreover, community conservation programmes themselves often consist of externally developed models that may prove of little local relevance, as at Okomu. Instead of applying templates, forest policy would benefit from flexible, responsive approaches that engage seriously with local ecology and political economy.

A key aspect of this, and this is the second implication of this study, is to build on already existing local conservation and land management practices. Drawing on centuries of local forest knowledge, these are far more attuned to local ecological conditions and economic needs than any externally imposed programmes. Yet in Edo State and elsewhere local management practices may today largely take the form of informal or illegal arrangements, with few formal opportunities for their development. The illegality of local practices, such as 'new method' plantain farming inside Okomu Reserve, has so far largely prevented recognition of their conservation and development potentials. But their current illegal status also makes their recognition all the more important, as it exposes local farmers to serious risks and insecurities. In Edo State they need protection and support less against state conservation projects than against other forms of development that may receive government sanction but are more environmentally destructive and potential sources of conflict. Vis-à-vis these forces communities and conservationists have shared interests. Recognising these, Okomu communities and conservationists have already formed alliances at key junctures, such as during the campaign to protect Iguowan Lake and the buffer zone around the wildlife sanctuary in 1994 and the more recent campaign involving the World Rainforest Movement and Friends of the Earth against Michelin's bulldozing of Iguobazuwa Reserve. Such alliances – increasingly emerging throughout the world in the face of the rapid expansion of international land grab, mining and resource extraction – can be strengthened and developed further by a greater recognition of the conservation value of existing local land management practices, be they 'legal' or not.

Notes

1. The relative scarcity of elephants may have been due to the West African ivory trade of the sixteenth and seventeenth centuries (see Ryder 1969).
2. See also http://nigeriaparkservice.org, viewed 27 July 2010.
3. I include this here as a key example of the innovation, adaptability and relative sustainability characteristic of many local practices, even though this 'new method' emerged only after the 1990s.
4. A wildlife sanctuary is defined by the African Convention on the Conservation of Nature and Natural Resources as an area 'set aside to protect characteristic wildlife … or to protect particularly threatened animal or plant species … together with the biotopes essential for their survival', in which 'all other interests and activities shall be subordinated to this end' (Oates and Anadu 1982: 31).
5. By means of Legal Notice No. 189 of 1986.
6. MANR, Department of Forestry, Benin City, FR 848T/229-232, Brief on the Proposed Technical Assistance of the ODA in the Management of Okomu Forest Reserve, 17 May 1991.
7. Ibid.
8. This is based on a conversation I had with a taxi driver from Nikrowa, who complained that the NCF and park never did anything for them, 7 January 2002.
9. By the National Parks Decree of 1999, Decree No. 46 of 1999.
10. These included Alfred Ohenhen and Okechukwu, who were both trained by the NCF as tourist guides. They continued to act as tourist guides under the national park management and were therefore allocated to me as research assistants when I conducted my fieldwork.
11. Yankari was handed over to Bauchi state government and ceased to be a national park in 2006.
12. See Cajetan Mmuta, *Compass Newspaper*, 10 March 2009, http://www.compassnewspaper.com/NG/index.php?option=com_content&view=article&id=12330%3Aexpert-cautions-edo-over-threats-to-okomu-national-park-&Itemid=7966, accessed 3 August 2010.

13. I witnessed this walking through this area with a park ranger. He collected a large amount of kernels used in soup, spread to dry outside the houses of the Yoruba camp, as compensation for keeping quiet about the fact that they were farming inside the national park.
14. In 2007, however, after my last visit, a female South African researcher based at Okomu national park was kidnapped by Ijaw youths.

Appendix

Administrative History of the Research Area

800–1500s: Overall part of emerging Benin Kingdom, but much of it contested, in particular the area west of the Ovia river, ruled by Udo.

1500s to 1897: Heartland of the Benin Kingdom.

1897–1900: After the conquest of Benin City, part of the Niger Coast Protectorate.

1900–1903: Part of the Protectorate of Southern Nigeria, formed through the union of the Niger Coast Protectorate and the territories of the Royal Niger Company.

1903–1906: Part of the Western Division of the Protectorate of Southern Nigeria.

1906–1914: Benin District of the Central Province of Southern Nigeria.

1914–1963: Benin Division of the Benin Province of Nigeria (Nigeria was formed through the amalgamation of Northern and Southern Nigeria in 1914; the area was then divided into provinces and further divided into divisions).

Note: In the 1930s and 1940s the Forest Department used slightly different administrative boundaries and names. The Benin Native Administration Forest Circle covered most but not all of the Benin Division. It was sometimes abbreviated to Benin Circle.

1963–1967: Part of the Midwest Region, formed through the amalgamation of the Benin Province and Delta Province.

1967–1976: Part of Midwest State (the same area as the Midwest Region).

1976–1991: Part of Bendel State (the same area as Midwest State)

1991–present: Part of Edo State, formed when Bendel State was split into Edo and Delta State.

Bibliography

Achebe, C. 1958. *Things Fall Apart*. Heineman, London.
Adams, C.J. 1966. *Remarks on the Country Extending from Cape Palmas to the River Congo with an Appendix Containing an Account of the European Trade with the West Coast of Africa*. Frank Cass, London.
Adams, P. 1995. 'Balance of Nature', *Financial Times*, 8 March.
Adams, W.M. and D. Hulme. 2001a. 'Conservation and Community. Changing Narratives, Policies and Practices in African Conservation', in *African Wildlife and Livelihoods. The Promise and Performance of Community Conservation*, edited by D. Hulme and M. Murphee. James Currey, Oxford.
———. 2001b. 'If Community Conservation Is the Answer in Africa, What Is the Question?', *Oryx* 35(3): 193–200.
Adeleke, A. 1999. 'Participatory Management Planning for Multi-Purpose Forest Resource Utilisation and Management in Okomu Forest Reserve, Edo State, Nigeria', MSc dissertation, University of Edinburgh.
Adeyoju, S.K. 1966. 'The Development of the Timber Industry in the Western and Midwestern Regions of Nigeria', Ph.D. dissertation, London School of Economics and Political Science.
Afigbo, A.E. 1970. 'Sir Ralph Moor and the Economic Development of Southern Nigeria: 1896–1903', *Journal of the Historical Society of Nigeria* 5(3): 371–397.
Agrawal, A. 2005. *Environmentality: Technologies of Government and the Making of Subjects*. Duke University Press, Durham, NC.
Ainslie, J.R. 1929. 'Annual Report on the Forest Administration of Nigeria, for the Year 1928', Forest Administration of Nigeria.
———. 1930. 'Annual Report on the Forest Administration of Nigeria for the Year 1929', Forest Administration of Nigeria.
Allison, P.A. 1941. 'From Farm to Forest', *Farm and Forest* 2(2): 95–98.
———. 1962. 'Historical Inferences to Be Drawn from the Effect of Human Settlement on the Vegetation of Africa', *The Journal of African History* 3(2): 241–249.
Aluko, J.O. 2006. *Corruption in the Local Government System in Nigeria*. BookBuilders, Ibadan.

Aluko, T.M. 1966. *Kinsman and Foreman*. Heinemann, London.
Amanor, K.S. 2001. 'The Symbolism of Tree Planting and Hegemonic Environmentalism in Ghana', paper presented at Changing Perspectives on Forests: Ecology, People and Science/Policy Processes in West Africa and the Caribbean, Institute of Development Studies, Sussex, March.
Andah, B. 1987. 'Agricultural Beginnings and Early Farming Communities in West and Central Africa', *West African Journal of Archaeology* 17: 171–204.
———. 1993. 'Identifying Early Farming Traditions of West Africa', in *The Archaeology of Africa. Food, Metals and Towns*, edited by T. Shaw, P. Sinclair, B. Andah and A. Okpoko, pp. 240–254. Routledge, London.
Anders, G. 2004. 'Like Chameleons. Civil Servants and Corruption in Malawi', *Le bulletin de l'APAD* 23–24: 43–67.
Anderson, D. 1984. 'Depression, Dust Bowl, Demography and Drought; the Colonial State and Soil Conservation in East Africa during the 1930s', *African Affairs* 83(332): 321–343.
———. 1987. 'Managing the Forest: The Conservation History of Lembus, Kenya, 1904–63', in *Conservation in Africa. People, Policies and Practice*, edited by D. Anderson and R. H. Grove, pp. 249–268. Cambridge University Press, Cambridge.
Anderson, D. and R.H. Grove (eds). 1987a. *Conservation in Africa. People, Policies and Practice*. Cambridge University Press, Cambridge.
———. 1987b. 'Introduction: The Scramble for Eden: Past, Present and Future in African Conservation', in *Conservation in Africa. People, Policies and Practice*, edited by D. Anderson and R.H. Grove, pp. 1–20. Cambridge University Press, Cambridge.
Anene, J.C. 1966. *Southern Nigeria in Transition 1885–1906. Theory and Practice in a Colonial Protectorate*. Cambridge University Press, Cambridge.
Anker, P. 2001. *Imperial Ecology. Environmental Order in the British Empire, 1895–1945*. Harvard University Press, Cambridge, MA.
Ashton-Jones, N. 1993. Memorandum by Nick Ashton-Jones, Africa Representative of Pro Natura International in West and Central Africa to the Nigerian Conservation Foundation, Scientific Committee Chairman. An Assessment of the Okomu Forest Project.
Aubréville, A. 1938. 'La Forêt Colonial: Les Forêts De L'afrique-Occidentale Française', *Annales d'Académie des Sciences Coloniales* IX: 1–245.
Balée, W. (ed.). 1998. *Advances in Historical Ecology*. Columbia University Press, New York.
———. 2006. 'The Research Program of Historical Ecology', *Annual Review of Anthropology* 35(1): 75–98.
Barber, R.J. 1985. 'Land Snails and Past Environment at the Igbo-Iwoto Esie Site, Southwestern Nigeria', *West African Journal of Archaeology* 15: 89–102.

Barbier, E.B., R. Damania and D. Léonard. 2005. 'Corruption, Trade and Resource Conversion', *Journal of Environmental Economics and Management* 50(2): 276–299.

Barton, G. 2001. 'Empire Forestry and the Origins of Environmentalism', *Journal of Historical Geography* 27(4): 529–552.

Bauer, P.T. 1948. *The Rubber Industry: A Study in Competition and Monopoly*. Harvard University Press, Cambridge, MA.

Bayart, J.-F. 1993. *The State in Africa: The Politics of the Belly*. Longman, London.

Bayart, J.-F., S. Ellis and B. Hibon. 1999. *The Criminalization of the State in Africa*. African Issues. James Currey, Oxford.

Becroft, J. 1841. 'Account of a Visit to the Capital of Benin, in the Delta of the Kwara or Niger, in the Year 1838', *Journal of the Royal Geographical Society* 11: 190–192.

Becroft, J. and R. Jamieson. 1841. 'On Benin and the Upper Course of the River Quorra or Niger', *Journal of the Royal Geographical Society* 11: 184–90.

Bee, O.J. 1991. 'The Tropical Rain Forest: Patterns of Exploitation and Trade', *Singapore Journal of Tropical Geography* 11(2): 117–142.

Beier, P. and R.F. Noss. 1998. 'Do Habitat Corridors Provide Connectivity?', *Conservation Biology* 12(6): 1241–1252.

Beinart, W. 1984. 'Soil Erosion, Conservationism and Ideas About Development: A Southern African Exploration, 1900–1960', *Journal of Southern African Studies* 11(1): 52–83.

Ben-Amos, P.G. 1995. *The Art of Benin*. 2nd ed. British Museum Press, London.

Ben-Amos, P.G. and J. Thornton. 2001. 'Civil War in the Kingdom of Benin, 1689–1721: Continuity or Political Change?', *Journal of African History* 42: 353–376.

Bennett, J.A. 2000. '"The Grievous Mistakes of the Vanikoro Concession": The Vanikoro Kauri Timber Company, Solomon Islands, 1926–1964', *Environment and History* 6: 317–347.

Berman, B. and J. Lonsdale. 1992. *Unhappy Valley. Conflict in Kenya & Africa*. James Currey, London.

Bernstein, H. and P. Woodhouse. 2001. 'Telling Environmental Change Like It Is? Reflections on a Study in Sub-Saharan Africa', *Journal of Agrarian Change* 1(2): 283–324.

Berry, N., O. Phillips, R. Ong and K. Hamer. 2008. 'Impacts of Selective Logging on Tree Diversity across a Rainforest Landscape: The Importance of Spatial Scale', *Landscape Ecology* 23(8): 915–929.

Berry, S. 1993. *No Condition Is Permanent. The Social Dynamics of Agrarian Change in Sub-Saharan Africa*. The University of Wisconsin Press, Wisconsin.

Biersack, A. 2006. 'Reimagining Political Ecology: Culture/Power/History/Nature', in *Reimagining Political Ecology*, edited by A. Biersack and J.B. Greenberg, pp. 3–40. Duke University Press, Durham, NC and London.

Biersack, A. and J.B. Greenberg. 2006. *Reimagining Political Ecology*. Duke University Press, Durham, NC and London.

Blackett, H. and E. Gardette. 2008. *Cross-Border Flows of Timber and Wood Products in West Africa*. European Commission/HTSPE Ltd.

Blaikie, P. and H. Brookfield. 1987. *Land Degradation and Society*. Methuen, London.

Blake, J.W. (ed.). 1942. *Europeans in West Africa, 1450–1560*, 2 vols. Hakluyt Society, London.

Blundo, G. and J.-P. Olivier de Sardan. 2006. *Everyday Corruption and the State: Citizens and Public Officials in Africa*. Zed, London.

Boisragon, A. 1897. *The Benin Massacre (by one of the survivors)*. Methuen, London.

Bondarenko, D.M. and P.M. Roese. 1999. 'Benin Prehistory. The Origin and Settling Down of the Edo', *Anthropos* 94: 542–552.

Bosman, W. 1967. *A New and Accurate Description of the Coast of Guinea*. Frank Cass, London.

Botkin, D.B. 1990. *Discordant Harmonies. A New Ecology for the Twenty-First Century*. Oxford University Press, Oxford.

Bradbury, R.E. 1957. *The Benin Kingdom and the Edo-Speaking Peoples of South-Western Nigeria*. Ethnographic Survey of Africa XIII. International African Institute, London.

———. 1973a. 'The Benin Village', in *Benin Studies*, edited by P. Morton-Williams, pp. 149–209. Oxford University Press, London.

———. 1973b. 'Continuities and Discontinuities in Pre-Colonial and Colonial Benin Politics', in *Benin Studies*, edited by P. Morton-Williams, pp.76–128. Oxford University Press, London.

———. 1973c. 'The Kingdom of Benin', in *Benin Studies*, edited by P. Morton-Williams, pp. 44–75. Oxford University Press, London.

———. 1973d. 'Patrimonialism and Gerontocracy in Benin Political Culture', in *Benin Studies*, edited by P. Morton-Williams, pp. 129–146. Oxford University Press, London.

———. 1973e. 'Chronological Problems in Benin History', in *Benin Studies*, edited by P. Morton-Williams, pp. 17-43. Oxford University Press, London.

Brockington, D. 2002. *Fortress Conservation. The Preservation of Mkomazi Game Reserve, Tanzania*. James Currey, Oxford.

Brooks, G.E. 1993. *Landlords and Strangers: Ecology, Society and Trade in Western Africa, 1000–1630*. Westview Press, Boulder, CO and Oxford.

Brooks, K., C.A. Scholz, J.W. King, J. Peck, J.T. Overpeck, J.M. Russell and P.Y.O. Amoako. 2005. 'Late-Quaternary Lowstands of Lake Bosumtwi, Ghana: Evidence from High-Resolution Seismic-Reflection and Sediment-Core Data', *Palaeogeography, Palaeoclimatology, Palaeoecology* 216(3–4): 235–249.

Brown, J.C. and M. Purcell. 2005. 'There's Nothing Inherent about Scale: Political Ecology, the Local Trap, and the Politics of Development in the Brazilian Amazon', *Geoforum* 36(5): 607–624.

Brun, S. 1983. 'Samuel Brun, des Wundartzet und Burgers zu Basel, Schiffarten', in *German Sources for West African History, 1599–1669*, edited by A. Jones. Franz Steiner Verlag, Wiesbaden.

Bryant, R. 1994. 'Shifting the Cultivator: The Politics of Teak Regeneration in Colonial Burma', *Modern Asian Studies* 28(2): 225–250.

Bryant, R.L. 1994. 'The Rise and Fall of Taungya Forestry. Social Forestry in Defence of the Empire', *The Ecologist* 24(1): 21–26.

———. 1997. *The Political Ecology of Forestry in Burma, 1824–1994.* Hurst & Company, London.

———. 1998. 'Rationalising Forest Use in British Burma 1856–1942', in *Nature and the Orient*, edited by R.H. Grove, V. Damodaran and S. Sangwan. Oxford University Press, New Delhi.

Bürgi, M., E.W.B. Russell and G. Motzkin. 2000. 'Effects of Postsettlement Human Activities on Forest Composition in North-Eastern United States: A Comparative Approach', *Journal of Biogeography* 27(5): 1123–1138.

Burgess, H.B. 1956. *Annual Report on the Forest Administration of the Western Region of Nigeria, 1st April, 1955, to 31st March, 1956.* Forest Administration of Nigeria.

———. 1957. *Annual Report on the Forest Administration of the Western Region of Nigeria, 1st April, 1956–31st March, 1957.* Forest Administration of Nigeria.

Butler, R.A. and W.F. Laurance. 2008. 'New Strategies for Conserving Tropical Forests', *Trends in Ecology and Evolution* 23(9): 469–472.

Cain, P.J. and A.G. Hopkins. 1993. *British Imperialism. Crisis and Deconstruction 1914–1990.* Longman, London.

Callahan, J.C. 1985. 'The Mahogany Empire of Ichabod T. Williams & Sons, 1838–1973', *Journal of Forest History* 29(3): 120–130.

Cannon, C.H., D.R. Peart and M. Leighton. 1998. 'Tree Species Diversity in Commercially Logged Bornean Rainforest', *Science* 281(5381): 1366–1368.

Chabal, P. and J.-P. Daloz. 1999. *Africa Works. Disorder as Political Instrument.* African Issues. James Currey, Oxford.

Chandran, M.D.S. 1998. 'Shifting Cultivation, Sacred Groves and Conflicts in Colonial Forest Policy in the Western Ghats', in *Nature and the Orient*, edited by R.H. Grove, V. Damodaran and S. Sangwan, pp. 674–707. Oxford University Press, Oxford.

Clarke-Butler-Cole, R.F. 1943. 'Forest Farming (Taungya Plantations) in Benin', *Farm and Forest* 4(3): 103–115.

Cleaver, K. 1992. 'Deforestation in the Western and Central African Forest: The Agricultural and Demographic Causes, and Some Solutions', in *Conservation of West and Central African Rainforests*, edited by K. Cleaver et al., pp. 65–78. World Bank Environment Paper, Vol. 1. World Bank, Washington DC.

Cline-Cole, R. 1996. 'Dryland Forestry. Manufacturing Forests and Farming Trees in Nigeria', in *The Lie of the Land. Challenging Received Wisdom* on the African Environment, edited by M. Leach and R. Mearns, pp. 122–139. James Currey (Oxford) and Heinemann (Portsmouth, NH), London.

———. 2000. 'Redefining Forestry Space and Threatening Livelihoods in Colonial Northern Nigeria', in *Contesting Forestry in West Africa*, edited by R. Cline-Cole, pp. 36–63. Ashgate, Aldershot.

Clutton-Brock, J. 1999. 'Letter to the Editor', *New Scientist Magazine* 163(2205): 54.

Coates, B., F. Dimowo, Q.B.O. Anthonio, I. Inahoro and S.S. Orhiere. 1993. 'Cost–Benefit Analysis of Land Use Alternatives in Okomu Forest Reserve, Edo State. Nigeria', Nigerian Conservation Foundation.

Colfer, C.J.P., G.R. Dahal and D. Capistrano. 2008. *Lessons from Forest Decentralization: Money, Justice and the Quest for Good Governance in Asia-Pacific*. Earthscan, London.

Collier, F.S. 1948. 'Forest Administration Plan, 1946–1955: Under a Ten Year Plan of Development for Nigeria', Forest Department of Nigeria.

———. 1950. 'Annual Report on the Forest Administration of Nigeria for the Year 1948-49', Forest Department of Nigeria.

———. 1951. 'Annual Report on the Forest Administration of Nigeria for the Year 1949-50', Forest Department of Nigeria.

Collins, W.B. 1945. 'Palm and Produce', *Farm and Forest* 6(2): 67–68.

Connah, G. 1975. *The Archaeology of Benin. Excavations and other Researches in and around Benin City, Nigeria*. Clarendon Press, Oxford.

———. 1987. *African Civilizations. Precolonial Cities and States in Tropical Africa: An Archaeological Perspective*. Cambridge University Press, Cambridge.

Conte, C.A. 2004. *Highland Sanctuary: Environmental History in Tanzania's Usambara Mountains*. Ohio University Press, Athens, OH.

Contreras-Hermosila, A. 2000. 'The Underlying Causes of Forest Decline', Centre for International Forestry Research. Copies available from Occasional Papers, No. 30.

Cooper, F. 2002. *Africa since 1940: The Past of the Present*. Cambridge University Press, Cambridge.

Corbridge, S. and S. Kumar. 2002. 'Community, Corruption, Landscape: Tales from the Tree Trade', *Political Geography* 21(6): 765–788.

Cowen, M.P. and R.W. Shenton. 1994. 'British Neo-Hegelian Idealism and Official Colonial Practice in Africa: The Oluwa Land Case of 1921', *The Journal of Imperial and Commonwealth History* 22(2): 217–250.

Crone, G.R. (ed.). 1937. *The Voyages of Cadamosto and other Documents on Western Africa in the Second Half of the Fifteenth Century*. The Hakluyt Society, London.

Crumley, C.L. (ed.). 1994. *Historical Ecology: Cultural Knowledge and Changing Landscapes*. School of American Research Press, Santa Fe.

Curtin, P.D. 1965. *The Image of Africa. British Ideas and Action, 1780–1850*. Macmillan, London.

D'Andrea, A.C., M. Klee and J. Casey. 2001. 'Archaeobotanical Evidence for Pearl Millet (Pennisetum Glaucum) in Sub-Saharan West Africa', *Antiquity* 75: 341–348.
Dapper, O. 1668. *Neukeurige Beschrijvinge Der Afrikaensche Gewesten*. Jacob van Meurs, Amsterdam.
———.1670. *Umbstaendliche Und Eigentliche Beschreibung Von Afrika*. Jacob van Meurs, Amsterdam.
Darling, P. 1977. 'A Change of Territory: Attempts to Trace More Than a Thousand Years of Population Movements by the Benin and Ishan Peoples in Southern Nigeria', in *African Historical Demography: Proceedings of a Seminar Held in the Centre of African Studies, University of Edinburgh, 29 and 30 April 1977*. Centre for African Studies, Edinburgh.
———. 1984. *Archaeology and History in Southern Nigeria. The Ancient Linear Earthworks of Benin and Ishan*. Cambridge Monographs in African Archaeology 11, British Archaeological Reports International Series 215. B.A.R., Oxford.
———. 1995. *Masterplan for Okomu Forest Reserve*. WWF/ODA.
Dauvergne, P. 1993. 'The Politics of Deforestation in Indonesia', *Pacific Affairs* 66(4): 497–518.
Davis, D.K. 2007. *Resurrecting the Granary of Rome: Environmental History and French Colonial Expansion in North Africa*. Ohio University Press, Athens, OH.
———. 2009. 'Historical Political Ecology: On the Importance of Looking Back to Move Forward', *Geoforum* 40(3): 285–286.
DeAngelis, D.L. and J.C. Waterhouse. 1987. 'Equilibrium and Nonequilibrium Concepts on Ecological Models', *Ecological Monographs* 57(1): 1–21.
De Marees, P. 1602. *Beschryvinge Ende Historische Verhael Vant Gout Koninckrijk Van Gunea*. Amsterdam.
Demeritt, D. 2002. 'What Is the "Social Construction of Nature"? A Typology and Sympathetic Critique', *Progress in Human Geography* 26(6): 767–790.
Denevan, W.M. 1992. 'The Pristine Myth: The Landscape of the Americas in 1492', *Annals of the Association of American Geographers* 82: 369–385.
Dennett, R.E. 1914. 'Annual Report on the Forest Administration of Southern Nigeria for the Year 1913', Forestry Department. Copies available from CAL PROF 5/4/51.
Donald, P.F. 2004. 'Biodiversity Impacts of Some Agricultural Commodity Production Systems', *Conservation Biology* 18(1): 17–38.
Dove, M.R. 1993. 'A Revisionist View of Tropical Deforestation and Development', *Environmental Conservation* 20: 17–24.
Dumett, R.E. 2001. 'Tropical Forests and West African Enterprise: The Early History of the Ghana Timber Trade', *African Economic History* 29: 79–116.
Eberlein, R. 2006. 'On the Road to the State's Perdition? Authority and Sovereignty in the Niger Delta, Nigeria', *Journal of Modern African Studies* 44(4): 573–596.

EC-FAO. 2003. 'Experience of Implementing National Forest Programmes in Nigeria', EC-FAO Partnership Programme (2002–2003).
Egbhuche, U.P., A. Adeleke, P. Coasts and S.S. Orhiere. 1995. 'The Role of Non-Governmental Organisations in Community Participation: Experiences of Okomu Forest Reserve, Benin, Edo State', *24th Annual Conference of the Forestry Association of Nigeria*. Kaduna, Nigeria.
Egboh, E.O. 1978. 'British Colonial Administration and the Legal Control of the Forests of Lagos Colony and Protectorate 1897–1902: An Example of Economic Control under Colonial Regime', *Journal of the Historical Society of Nigeria* 9(3): 70–90.
———. 1985. *Forestry Policy in Nigeria, 1897–1960*. University of Nigeria Press, Nsukka.
Egbon, T.K. 1990. 'Okomu Oil Palm Company and Its Impact on the Socio-Economic Development of Ovia Local Government Area', unpublished paper, University of Benin.
Egharevba, J.U. 1968. *A Short History of Benin*, 4th ed. Ibadan University Press, Ibadan.
Ekeh, P.P. 1975. 'Colonialism and the Two Publics in Africa: A Theoretical Statement', *Comparative Studies in Society and History* 17(1): 91–112.
Ekpo, M. 1979. 'Gift-Giving Practices and Bureaucratic Corruption in Nigeria', in *Bureaucratic Corruption in Sub-Saharan Africa: Toward a Search for Causes and Consequences*, edited by M. Ekpo. University Press of America, Washington DC.
Ekundare, R.O. 1973. *An Economic History of Nigeria, 1860–1960*. Methuen, London.
Elias, T.O. 1951. *Nigerian Land Law and Custom*. Routledge & Kegan Paul, London.
Enabor, E.E., J. A. Okojie and I. Verinumbe. 1982. 'Taungya Systems from Biological and Production Viewpoints', in *Agro-Forestry in the African Humid Tropics. Proceedings of a Workshop Held in Ibadan, Nigeria, 27 April–1 May 1981*.
Enahoro, P. 1966. *How to Be a Nigerian*. Caxton Press, West Africa.
Enogholase, G. 2007. 'Sale of Forest Reserve Causes Tension in Edo', *The Vanguard*, Lagos, 23 July.
Escobar, A. 1999. 'After Nature. Steps to an Antiessentialist Political Ecology', *Current Anthropology* 40(1): 1–30.
Eweka, P.E.B. 1992. *Evolution of Benin Chieftaincy Titles*. Uniben Press, Benin City.
Ezele, C.J.I. 2002a. 'A Brief History of Bini Traditional Marriage', unpublished manuscript, Udo.
———. 2002b. 'A Short History of Udo', unpublished manuscript, Udo.
Fairhead, J. and M. Leach. 1996. *Misreading the African Landscape. Society and Ecology in a Forest-Savanna Mosaic*. Cambridge University Press, Cambridge.
———. 1998. *Reframing Deforestation. Global Analysis and Local Realities: Studies in West Africa*. Routledge, London.

———. 2003. *Science, Society and Power. Environmental Knowledge and Policy in West Africa and the Caribbean*. Cambridge University Press, Cambridge.
Falola, T. 2004. *Economic Reforms and Modernization in Nigeria: 1945–1965*. Kent State University Press, Kent, OH.
Fawckner, C.J. 1837. *Narrative of Captain James Fawkner's Travels on the Coast of Benin*. A. Schloss, London.
Fenske, J. 2011. WP/29 Trees, Tenure and Conflict: Rubber in Colonial Benin. Wider Working Paper, UNU-WIDER.
Fitzherbert, E.B., M.J. Struebig, A. Morel, F. Danielsen, C.A. Brühl, P.F. Donald and B. Phalan. 2008. 'How Will Oil Palm Expansion Affect Biodiversity?', *Trends in Ecology and Evolution* 23(10): 538–545.
Forestry Department, Edo State. 2002a. 'Brief on Forestry Activities in Edo State by the Director of Forestry Presented to the Honourable Commissioner for Agriculture and Natural Resources on the 16th May 2002', Department of Forestry, Ministry of Agriculture and Natural Resources.
———. 2002b. 'Forest Reserves in Edo State Showing Areas De-Reserved, Internal Communication', Department of Forestry, Ministry of Agriculture and Natural Resources.
Forrest, T. 1993. *Politics and Economic Development in Nigeria*. African Modernization and Development Studies. Westview Press, Oxford.
Foucault, M. 2000. 'Governmentality', in *Power. Essential Works of Foucault 1954–1984*, Vol. 3, edited by J.D. Faubian. Penguin Books, London.
Francis, P. 1984. '"For the Use and Common Benefit of All Nigerians": Consequences of the 1978 Land Nationalization', *Africa* 54(3): 5–27.
Franzen, M. and M.B. Mulder. 2007. 'Ecological, Economic and Social Perspectives on Cocoa Production Worldwide', *Biodiversity and Conservation* 16(13): 3835–3849.
Freund, B. 1998. *The Making of Contemporary Africa*. Macmillan Press, London.
Gallwey, H. 1893. 'Journeys in the Benin Country, West Africa', *Geographical Journal* 1: 122–130.
Geist, H.J. and E.F. Lambin. 2002. 'Proximate Causes and Underlying Driving Forces of Tropical Deforestation', *Bioscience* 52: 143–150.
German, L., A. Karsenty and A.-M. Tiani (eds). 2010. *Governing Africa's Forests in a Globalized World*. Earthscan, London.
Germer, J. and J. Sauerborn. 2008. 'Estimation of the Impact of Oil Palm Plantation Establishment on Greenhouse Gas Balance', *Environment, Development and Sustainability* 10(6): 697–716.
Giles-Vernick, T. 2002. *Cutting the Vines of the Past. Environmental Histories of the Central African Rain Forest*. University Press of Virginia, Charlottesville.
Gillson, L., M. Sheridan and D. Brockington. 2003. 'Representing Environments in Flux: Case Studies from East Africa', *Area* 35(4): 371–389.
Glastra, R. 1999. *Cut and Run: Illegal Logging and Timber Trade in the Tropics*. International Development Research Centre, Ottawa.

Gonggryp, J.W. 1948. 'Outline of a General Forest Policy for the Tropics', *Unasylva* 2(1): 3–7.
Gore, C. and D. Pratten. 2003. 'The Politics of Plunder: The Rhetorics of Order and Disorder in Southern Nigeria', *African Affairs* 102(407): 211–240.
Graham, J.D. 1965. 'The Slave Trade, Depopulation and Human Sacrifice in Benin History', *Cahiers d'Etudes Africaines* 5(2): 317–334.
Grainger, A. and W. Konteh. 2007. 'Autonomy, Ambiguity and Symbolism in African Politics: The Development of Forest Policy in Sierra Leone', *Land Use Policy* 24: 42–67.
Grove, R.H. 1990. 'Colonial Conservation, Ecological Hegemony and Popular Resistance: Towards a Global Synthesis', in *Imperialism and the Natural World*, edited by J.M. MacKenzie. Manchester University Press, Manchester.
———. 1995. *Green Imperialism. Colonial Expansion, Tropical Island Edens and the Origins of Environmentalism, 1600–1860*. Studies in Environment and History. Cambridge University Press, Cambridge.
———. 1997. 'Chiefs, Boundaries and Sacred Woodlands: Early Nationalism and the Defeat of Colonial Conservationism in the Gold Coast and Nigeria, 1870–1916', in *Ecology, Climate and Empire. Colonialism and Global Environmental History, 1400–1940*, edited by R.H. Grove. The White Horse Press, Cambridge.
Groves, A. 2008. *Shell and Society: Securing the Niger Delta: (Un)Civil Society and Corporate Security Strategies in the Niger Delta*, MSc dissertation, University of Oxford.
Guha, R. 1989. *The Unquiet Woods: The Ecological Basis of Peasant Resistance in the Himalaya*. Oxford University Press, New Delhi.
Guyer, J. and P. Richards. 1996. 'The Invention of Biodiversity: Social Perspectives on the Management of Biological Variety in Africa', *Africa* 66(1): 1–13.
Guyer, J.I. and S.M.E. Belinga. 1995. 'Wealth in People as Wealth in Knowledge: Accumulation and Composition in Equatorial Africa', *The Journal of African History* 36(1): 91–120.
Hagberg, S. 2002. 'Enough Is Enough: An Ethnography of the Struggle against Impunity in Burkina Faso', *The Journal of Modern African Studies* 40(2): 217–246.
Hakluyt, R. (ed.). 1907. *The Principal Navigations, Voyages, Traffiques, and Discoveries of the British Nation*, Vol. 4. J.M Dent and Sons, London.
Hamman, D.A. and G.P.E. Stopfords. 2001. 'National Park Service Goes Paramilitary', *Nigeria Parks. The Magazine of the Nigeria National Parks* 2: 26–27.
Harrison, E. 2006. 'Unpacking the Anti-Corruption Agenda: Dilemmas for Anthropologists', *Oxford Development Studies* 34(1): 15–29.
Hartemink, A. 2005. 'Plantation Agriculture in the Tropics. Environmental Issues', *Outlook on Agriculture* 34(1): 11–21.

Hastings, A., C.L. Hom, S. Ellner, S. Turchin and H.C. Godfray. 1993. 'Chaos in Ecology: Is Mother Nature a Strange Attractor?', *Annual Review of Ecology and Systematics* 24: 1–33.

Haworth, J. 1999. *Life after Logging. The Impact of Commercial Timber Extraction in Tropical Rainforests*. Friends of the Earth Trust/The Rainforest Foundation.

Hawthorne, W.D. 1995. 'Ecological Profiles of Ghanaian Forest Trees', *Tropical Forestry Papers No. 29*. Oxford Forestry Institute, Oxford.

———. 1996. 'Holes and the Sums of Parts in Ghanaian Forest: Regeneration, Scale and Sustainable Use', *Proceedings of the Royal Society of Edinburgh* 104B: 75–176.

Hecht, S. and A. Cockburn. 1990. *The Fate of the Forest*. Penguin, London.

Helleiner, G.K. 1966. *Peasant Agriculture, Government, and Economic Growth in Nigeria*. Richard D. Irwin, Homewood, Illinois.

Henige, D. 1987. 'The Race is not always to the Swift. Thoughts on the Use of Written Sources for the Study of Early African History', *Paideuma* 33: 53–79.

Henry, P.W.T. 1956. 'Development of the Regeneration Obtained by the Tropical Shelterwood System when Applied as a Pre-Exploitation Treatment Only', *Nigerian Forests Information Bulletin* 40: 1–3.

Heussler, R. 1963.*Yesterday's Rulers. The Making of the British Colonial Service*. Syracuse University Press, New York.

Hide, R.H. 1943. 'The Bini as a Botanist', *The Nigerian Field* 9: 169–179.

Igbafe, P.A. 1979. *Benin under British Administration. The Impact of Colonial Rule on an African Kingdom 1897–1938*. Ibadan History Series. Longman, London.

———. 1980. 'The Pre-Colonial Economic Foundations of Benin', in *Topics of Nigerian Economic and Social History*, edited by I.A. Akinjogbin. Ife History Series, edited by I.A. Akinjogbin. University of Ife Press, Ile-Ife.

Imafidon, A. 1987. 'Edo-Udo Relations: An Aspect of the External Relations of the Benin Kingdom', MA dissertation, University of Benin.

Ingold, T. and J.L. Vergunst. 2008. *Ways of Walking: Ethnography and Practice on Foot*. Ashgate, Aldershot.

Inikori, J.E. 1996. 'Slavery in Africa and the Transatlantic Slavetrade', in *The African Diaspora*, edited by A. Jalloh and S.E. Maizlish, pp. 39–71. Texas University Press, Arlington.

Ite, U.E. 1998. 'New Wine in an Old Skin: The Reality of Tropical Moist Forest Conservation in Nigeria', *Land Use Policy* 15(2): 135–147.

———. 2000. 'Assessing Conservation-with-Development in Cross River National Park, Nigeria', in *Contesting Forestry in West Africa*, edited by R. Cline-Cole and C. Madge, pp. 178–200. The Making of Modern Africa, edited by A. Zegeye and J. Higginson. Ashgate, Aldershot.

Johnson, C.O. 2003. 'Nigeria: Illegal Logging and Forest Women's Resistance', *Review of African Political Economy* 30(95): 156–162.

Jones, A. (ed.) 1983. *German Sources for West African History, 1599–1669*. Franz Steiner Verlag, Wiesbaden.

———. 1987. 'A Critique of Editorial and Quasi-Editorial Work on Pre-1885 European Sources for Sub-Saharan Africa, 1960–1986', *Paideuma* 33: 95–105.

———. 1998. *Olfert Dapper's Description of Benin (1668)*. University of Wisconsin, Madison.

Jones, A. and B. Heintze. 1987. 'European Sources for Sub-Saharan African before 1900: Use and Abuse', *Paideuma* 33: special edition

Jones, E.W. 1955. 'Ecological Studies on the Rain Forest of Southern Nigeria: Iv (Continued) the Plateau Forest of the Okomu Forest Reserve, Part 1. The Environment, the Vegetation Types of the Forest, and the Horizontal Distribution of Species', *Journal of Ecology* 43: 564–594.

———. 1956. 'Ecological Studies on the Rain Forest of Southern Nigeria: Iv (Continued) the Plateau Forest of the Okomu Forest Reserve, Part 2. The Reproduction and History of the Forest', *Journal of Ecology* 44: 83–117.

Joseph, R.A. 1983. 'Class, State and Prebendal Politics in Nigeria', *Journal of Commonwealth and Comparative Politics* 21(3): 21–38.

Keay, R.W.J. 1947. 'Notes on the Vegetation of Old Oyo Forest Reserve', *Farm and Forest* 8(1): 36–46.

Keeley, J. and I. Scoones. 2003. *Understanding Environmental Policy Processes. Cases from Africa*. Earthscan, London.

Kelsall, T. 2003. 'Rituals of Verification: Indigenous and Imported Accountability in Northern Tanzania', *Africa* 73(2): 174–201.

Kennedy, J.D. 1930. 'Taungya Method of Regeneration in Nigeria', *Empire Forestry Journal* 9(2): 221–225.

King-Church, L.A. 1920. 'Annual Report on the Forestry Department, Southern Provinces, for the Year 1919', Forestry Department of Nigeria.

Kio, P.R.O. 1978. *Stand Development in Naturally Regenerated Forest in South-Western Nigeria*, PhD dissertation, University of Ibadan.

Kirk-Greene, A. and D. Rimmer. 1981. *Nigerian since 1970. A Political and Economic Outline*. Hodder and Stoughton, London.

Kishor, N. and R. Damania. 2007. 'Crime and Justice in the Garden of Eden: Improving Governance and Reducing Corruption in the Forestry Sector', in *The Many Faces of Corruption: Tracking Vulnerabilities at the Sector Level*, edited by J.E. Campos and S. Pradhan, pp. 89–114. The World Bank, Washington DC.

Klopp, J.M. 2000. 'Pilfering the Public: The Problem of Land Grabbing in Contemporary Kenya', *Africa Today* 47(1): 7–26.

Lamb, A.F.A. 1966. 'Impressions of Nigerian Forestry after an Absence of Twenty-Three Years', unpublished manuscript, Commonwealth Forestry Institute.

Landolphe, C.J.F. 1823. *Memoires Du Capitaine Landolphe, Contenant L'histoire De Ses Voyages*, 2 vols. A. Bertrand, Paris.

Larson, A.M. and J.C. Ribot. 2007. 'The Poverty of Forestry Policy: Double Standards on an Uneven Playing Field', *Sustainability Science* 2(2): 189–204.
Laurance, W.F. 2004. 'The Perils of Payoff: Corruption as a Threat to Global Biodiversity', *Trends in Ecology and Evolution* 19(8): 399–401.
Leach, M. and R. Mearns (eds).1996. *The Lie of the Land. Challenging Received Wisdom on the African Environment*. James Currey, Oxford.
Letouzey, R. 1968. *Etude Phytogeographique Du Cameroun*. P. Lechavelier, Paris.
Lewis, P. 1996. 'From Prebendalism to Predation: The Political Economy of Decline in Nigeria', *The Journal of Modern African Studies* 34(1): 79–103.
Lipsky, M. 1979. *Street Level Bureaucracy*. Russell Sage Foundation, New York.
Lloyd, P.C. 1991. 'Report on a Socio-Economic Survey in the Okomu Forest Reserve to the Government of Nigeria Bendel State Forestry Project', University of Sussex, under assignment from the Overseas Development Administration.
Lloyd, P.C. and F. Omorodion. 1992. 'Socio Economic Survey of the Okomu Forest Reserve Edo State, Nigeria', University of Sussex, University of Benin.
Lowe, R.G. 1987. 'Development of Taungya in Nigeria', in *Agroforestry: Realities, Possibilities and Potentials*, edited by H.L. Gholz. Martinus Nijhoff Publishers, in cooperation with ICRAF.
———. 1994. 'Silviculture in Moist Lowland Forests of West and Central Africa. A Review of Its Development, Together with Recommendations for Successful Establishment of Forestry Plantations', The Government of Cameroon, Ministry of Environment and Forestry.
———. 2000. 'Forestry in Nigeria: Past, Present and Future', *The Nigerian Field* 65: 58–71.
———. 2003. 'Traditional Taungya in Africa', *Quarterly Journal of Forestry* 97(1): 25–28.
Lowe, R.G., J. Caldecott, R. Barnwell and R.W.J. Keay. 1992. 'Nigeria', in *Conservation Atlas of Tropical Forests: Africa*, edited by J. Sayer, C.S. Harcourt and N.M. Collins, pp. 230–239. Macmillan, London.
Lowood, H.E. 1990. 'The Calculating Forester: Quantification, Cameral Science, and the Emergence of Scientific Forestry Management in Germany', in *The Quantifying Spirit in the 18th Century*, edited by T. Frängsmyr, J.L. Heilbron and R.E. Rider, pp. 315–342. University of California Press, Berkeley.
Lugard, L. 1970. *Political Memoranda. Revision of Instructions to Political Officers on Subjects Chiefly Political and Administrative, 1913–1918*, 3rd ed. Frank Cass, London.
MacKenzie, J.M. 1989. *The Empire of Nature: Hunting, Conservation and British Imperialism*. University of Manchester Press, Manchester.
Maier, K. 2000. *This House Has Fallen. Nigeria in Crisis*. Allen Lane, London.
Maley, J. 1996. 'Fluctuations Majeures De La Forêt Dense Humide Africaine', in *L'alimentation En Forêt Tropicale*, Vol. 1, edited by C.M. Hladik, A. Hladik, H. Pagezy, O.F. Linares, G.J.A. Koppert and A. Froment. Editions UNESCO.

———. 2001. 'Elaeis Guineensis Jacq. (Oil Palm) Fluctuations in Central Africa during the Late Holocene: Climate or Human Driving Forces for This Pioneering Species?', *Vegetation History and Archaeobotany* 10(2): 117–120.

———. 2002. 'A Catastrophic Destruction of African Forests about 2,500 Years ago still Exerts a Major Influence on Present Vegetation Formations', *IDS Bulletin* 33(1): 13–30.

Mallam, J.C. 1953. 'Regeneration of High Forest', *Nigerian Forests Information Bulletin* 11(1): 1.

Malleson, R. 1999. 'Forest Livelihoods in Southwest Province, Cameroon: An Evaluation of the Korup Experience', Ph.D. dissertation, University College London.

———. 2002. 'Changing Perspectives on Forests, People and "Development": Reflections on the Case of the Korup Forest', *IDS Bulletin* 33(1): 94–101.

Mamdani, M. 1996. *Citizen and Subject. Contemporary Africa and the Legacy of Late Colonialism*. James Currey, London.

Marguba, L.B. 2000. 'Development of National Parks in Nigeria – an Overview', *Nigeria Parks. The Magazine of the Nigeria National Parks* 2: 1–3.

Marshall, H.F. 1939. *Intelligence Report on the Udo and Siluko Districts*, Secretary's Office, Western Provinces.

Martin, L. 1997. *Commerce and Economic Change in West Africa: The Palm Oil Trade in the Nineteenth Century*. Cambridge University Press, Cambridge.

Martin, S.M. 1988. *Palm Oil and Protest. An Economic History of the Ngwa Region, South-Eastern Nigeria, 1800–1980*. African Studies, Series 59. Cambridge University Press, Cambridge.

McCann, J.C.1999. *Green Land, Brown Land, Black Land. An Environmental History of Africa, 1800–1990*. James Currey, Oxford.

McEwan, C. 2000. 'Representing West African Forests in British Imperial Discourse C. 1830–1900', in *Contesting Forestry in West Africa*, edited by R. Cline-Cole and C. Madge, pp. 16–35. The Making of Modern Africa, edited by A. Zegeye and J. Higginson. Ashgate, Aldershot.

McLeod, N.C. 1908. 'Report on the Forest Administration of Southern Nigeria, 1907', Forestry Department of Southern Nigeria.

Meagher, K. 2007. 'Hijacking Civil Society: The Inside Story of the Bakassi Boys Vigilante Group of South-Eastern Nigeria', *Journal of Modern African Studies* 45(1): 89–115.

Meek, C.K. 1937. *Law and Authority in a Nigerian Tribe. A Study in Indirect Rule*. Geoffrey Cumberledge/Oxford University Press, London.

Melville, R. 1936. 'A List of True and False Mahoganies', *Bulletin of Miscellaneous Information (Royal Gardens, Kew)* 1936(3): 193–210.

Melzian, H. 1937. *A Concise Dictionary of the Bini Language of Southern Nigeria*. Kegan Paul, Trench, Trubner & Co, London.

Milligan, S. and T. Binns. 2007. 'Crisis in Policy, Policy in Crisis: Understanding Environmental Discourse and Resource-Use Conflict in Northern Nigeria', *Geographical Journal* 173(2): 143–56.

Moloney, A. 1887. *Sketch of the Forestry of West Africa, with Particular Reference to Its Present Principal Commercial Products*. Sampson Low, Marston, Searle, & Rivington, London.

Morgan, W.B. 1959. 'The Influence of European Contacts on the Landscape of Southern Nigeria', *Geographical Journal* 125(1): 48–64.

Mosse, D. 2004. *Cultivating Development: An Ethnography of Aid Policy and Practice*. Pluto, London.

Munyae, M.M. and M.M. Mulinge. 1999. 'The Centrality of a Historical Perspective to the Analysis of Modern Social Problems in Sub-Saharan Africa: A Tale from Two Case Studies', *Journal of Social Development in Africa* 14(2): 51–70.

Mustapha, A.R. 2002. 'States, Predation and Violence: Reconceptualizing Political Action and Political Community in Africa', *10th General Assembly of CODESRIA*, Kampala, Uganda.

Mutch, W.E.S. 1952. 'The Tropical Shelterwood System of Forest Regeneration. Its Development and Application in the Benin Division of Southern Nigeria and a Consideration of Factors Affecting Its Success', Ph.D. dissertation, University of Edinburgh.

Neumann, K. 2006. 'Ölpalme, Perlhirse und Banane. Wie Kam Die Landwirtschaft in Den Regenwald Zentralafrikas?', *Forschung Frankfurt* 24(2–3): 38–41.

Neumann, R.P. 1997. 'Primitive Ideas: Protected Area Buffer Zones and the Politics of Land in Africa', *Development and Change* 28(3): 559–582.

———. 1998. *Imposing Wilderness. Struggles over Livelihood and Nature Preservation in Africa*. University of California Press, Berkeley.

Ngomanda, A., K. Neumann, A. Schweizer and J. Maley. 2009. 'Seasonality Change and the Third Millennium BP Rainforest Crisis in Southern Cameroon (Central Africa)', *Quaternary Research* 71(3): 307–318.

Nicholson, S.E. 1978. 'Climatic Variations in the Sahel and Other African Regions during the Past Five Centuries', *Journal of Arid Environments* 1: 3–24.

———. 1980. 'Saharan Climates in Historic Times', in *The Sahara and the Nile: Quaternary Environments and Prehistoric Occupations in Northern Africa*, edited by M.A.J. Williams and H. Faure, pp. 173–200. A.A. Balkema, Rotterdam.

Nicolson, I.F. 1969. *The Administration of Nigeria, 1900–1960. Men, Methods and Myths*. Clarendon Press, Oxford.

Njoku, U.I. 2005. 'Colonial Political Re-Engineering and the Genesis of Modern Corruption in African Public Service: The Issue of Warrant Chiefs in South Eastern Nigeria as a Case in Point', *Nordic Journal of African Studies* 14(1): 99–116.

Oates, J.F. 1995. 'The Danger of Conservation by Rural Development – A Case-Study from the Forests of Nigeria', *Oryx* 29(2): 115–122.

———. 1999. *Myth and Reality in the Rain Forest. How Conservation Strategies Are Failing in West Africa*. University of California Press, London.

Oates, J.F. and P.A. Anadu. 1982. 'The Status of Wildlife in Bendel State, Nigeria, with Recommendations for Its Conservation', WWF/IUCN Project No. 1613, unpublished report submitted to Nigerian Government, NYZS and WWF-US.

Ohadike, D.C. 1992. 'Igbo-Benin Wars', in *Warfare and Diplomacy in Pre-Colonial Nigeria: Essays in Honour of Robert Smith*, edited by T. Falola and R. Law. African Studies Program, University of Wisconsin, Madison.

Okali, D.U.U. and B.A. Ola-Adams. 1987. 'Tree Population Changes in Treated Rain Forest at Omo Forest Reserve, South-Western Nigeria', *Journal of Tropical Ecology* 3(4): 291–313.

Okpewko, I. 1998. *Once Upon a Kingdom. Myth, Hegemony and Identity*. Indiana University Press, Bloomington & Indianapolis.

Okuneye, P.A. 2002. 'Rising Cost of Food Prices and Food Insecurity in Nigeria and Its Implications for Poverty Reduction', *Central Bank of Nigeria Economic and Financial Review* 39(4).

Olaleye, O.A. and C.E. Ameh. 1999. *Forest Resource Situation Assessment of Nigeria*. FAO/European Commission, Directorate-Generale VIII Development. Copies available from PROJECT GCP/INT/679/EC.

Oliphant, J.N. 1940. 'Annual Report on the Forest Administration of Nigeria for the Year 1939', Forest Administration of Nigeria.

———. 1941. 'Annual Report on the Forest Administration of Nigeria for the Year 1940', Forest Administration of Nigeria.

Olivier de Sardan, J.P. 1999. 'A Moral Economy of Corruption in Africa?', *The Journal of Modern African Studies* 37(1): 25–52.

Omorodion, F. 1991. 'Socio-Economic Survey of Okomu Forest Reserve. A Report Submitted to the Nigerian Conservation Foundation', Department of Sociology and Anthropology, University of Benin.

Omosini, O. 1978. 'Background to the Forestry Legislation in Lagos Colony and Protectorate, 1897–1902', *Journal of the Historical Society of Nigeria* 9(3): 45–69.

Osaghae, E.E. 1998. *Crippled Giant. Nigeria since Independence*. Hurst & Company, London.

Osemwota, O. 1989. 'Ownership, Control and Management of Land in Bendel State, Nigeria: The Changing Role of Traditional Rulers', *Land Use Policy* 6(1): 75–83.

Osoba, S.O. 1996. 'Corruption in Nigeria: Historical Perspectives', *Review of African Political Economy* 23(69): 371–386.

Palmer, C.E. 2001. 'The Extent and Causes of Illegal Logging: An Analysis of a Major Cause of Tropical Deforestation in Indonesia', CSERG Working Paper.

Parren, M. 2003. *Lianas and Logging in West Africa*. Tropenbos – Cameroon Series 6. Tropenbos International, Wageningen.

Peet, R. and M. Watts. 2004. *Liberation Ecologies. Environment, Development, Social Movements*, 2nd ed. Routledge, London.

Peluso, N.L. 1992. *Rich Forests, Poor People: Resource Control and Resistance in Java*. University of California Press, Berkeley.

Pereira, D.P. 1937. *Esmeraldo De Situ Orbis*, translated by G.H.T. Kimble. Hakluyt Society, London.
Phillips, A. 1989. *The Enigma of Colonialism. British Policy in West Africa*. James Currey, London.
Pickett, S.T.A. and P.S. White (eds). 1985. *The Ecology of Natural Disturbance and Patch Dynamics*. Academic Press, Orlando.
Pierson, P. 2004. *Politics in Time: History, Institutions, and Social Analysis*. Princeton University Press, Princeton.
Plumptre, A.J. 1996. 'Changes Following 60 Years of Selective Timber Harvesting in the Budongo Forest Reserve, Uganda', *Forest Ecology and Management* 89(1–3): 101–113.
Poorter, L., F. Bongers, R.S.A.R. van Rompaey and M. de Klerk. 1996. 'Regeneration of Canopy Tree Species at Five Sites in West African Moist Forest', *Forest Ecology and Management* 84(1–3): 61–69.
Powell, J.M. 2007. 'Dominion over Palm and Pine: The British Empire Forestry Conferences, 1920–1947', *Journal of Historical Geography* 33(4): 852–877.
Pratt, M.L. 1992. *Imperial Eyes: Travel Writing and Transculturation*. Routledge, London.
Pratten, D. 2007. 'The Politics of Vigilance in Southeastern Nigeria', in *Twilight Institutions: Public Authority and Local Politics in Africa*, edited by C. Lund. Blackwell Publishing, Oxford.
Rajan, R. 1998. 'Imperial Environmentalism or Environmental Imperialism? European Forestry, Colonial Foresters and the Agendas of Forest Management in British India 1800–1900', in *Nature and the Orient*, edited by R.H. Grove, V. Damodaran and S. Sangwan, pp. 324–372. Oxford University Press, Oxford.
———. 2006. *Modernizing Nature: Forestry and Imperial Eco-Development, 1800–1950*. Oxford University Press, Oxford.
Rangarajan, M. 1998. 'Production, Desiccation and Forest Management in the Central Provinces 1850–1930', in *Nature and the Orient*, edited by R.H. Grove, V. Damodaran and S. Sangwan, pp. 575–595. Oxford University Press, Oxford.
Ranger, T. 1983. 'The Invention of Tradition in Colonial Africa', in *The Invention of Tradition*, edited by E. Hobsbawm and T. Ranger. Cambridge University Press, Cambridge.
Rappaport, R.A. 1967. *Pigs for the Ancestors; Ritual in the Ecology of a New Guinea People*. Yale University Press, New Haven.
Redhead, J.F. 1992. 'The Forest Kingdom of Benin, Nigeria', *The Nigerian Field* 57(3–4): 113–118.
Reno, W. 2002. 'The Politics of Insurgency in Collapsing States', *Development and Change* 33(5): 837–858.
Ribot, J.C. and A.M. Larson. 2005. *Democratic Decentralisation through a Natural Resource Lens*. Routledge, London.
Rice, R.A. and R. Greenberg. 2000. 'Cacao Cultivation and the Conservation of Biological Diversity', *AMBIO: A Journal of the Human Environment* 29(3): 167–173.

Richards, P.W. 1952. *The Tropical Rain Forest*, 1st ed. Cambridge University Press, Cambridge.

Robbins, P. 2000. 'The Rotten Institution: Corruption in Natural Resource Management', *Political Geography* 19(4): 423–443.

———. 2004. *Political Ecology: A Critical Introduction*. Critical Introductions to Geography. Blackwell Publishing, Malden, MA and Oxford.

Robbins, P., A.K. Chhangani, J. Rice, E. Trigosa and S M. Mohnot. 2007. 'Enforcement Authority and Vegetation Change at Kumbhalgarh Wildlife Sanctuary, Rajastan, India', *Environmental Management* 40: 365–378.

Robbins, P., K. McSweeney, A.K. Chhangani and J. Rice. 2009. 'Conservation as It Is: Illicit Resource Use in a Wildlife Reserve in India', *Human Ecology* 37: 559–575.

Robbins, P., K. McSweeney, T. Waite and J. Rice. 2006. 'Even Conservation Rules Are Made to Be Broken: Implications for Biodiversity', *Environmental Management* 37: 2162–2169.

Robson, M. 1955. 'Annual Report on the Forest Administration of the Western Region of Nigeria, 1st April, 1953 to 31st March, 1954', Government Printer/Western Region of Nigeria.

Rochel, X. 2005. 'The 18th Century Paintings of Raon L'etape: A Geo-Historical Interpretation', paper presented at the conference *History and Sustainability: Third International Conference of the European Society for Environmental History*, Florence, February.

Roe, E.M. 1995. 'Except-Africa: Postscript to a Special Section on Development Narratives', *World Development* 23(6): 1065–1069.

Roese, P.M. and D.M. Bondarenko. 2003. *A Popular History of Benin: The Rise and Fall of a Mighty Forest Kingdom*. Peter Lang, Frankfurt am Main and Oxford.

Rosevear, D.R. 1952a. 'Annual Report on the Forest Administration of Nigeria for the Year 1950–1951', Forestry Department of Nigeria.

———. 1952b. 'Introduction', *Nigerian Forests Information Bulletin* 1: 1–3.

———. 1952c. 'Policy – How Much Forest Is Necessary?' *Nigerian Forests Information Bulletin* (4): 1–5.

———. 1953. 'Annual Report on the Forest Administration of Nigeria 1951–1952', Forestry Department of Nigeria.

———. 1954. 'Annual Report on the Forest Administration of Nigeria for the Year 1952–1953', Forestry Department of Nigeria.

Roth, H.L. 1968. *Great Benin. Its Customs, Art and Horrors*. Routledge & Kegan Paul, London.

Rudel, T.K. 2007. 'Changing Agents of Deforestation: From State-Initiated to Enterprise Driven Processes, 1970–2000', *Land Use Policy* 24(1): 35–41.

Ryder, A.F.C. 1969. *Benin and the Europeans 1485–1897*. Ibadan History Series. Longmans, Green and Co., London.

Salzmann, U. and P. Hoelzmann. 2005. 'The Dahomey Gap: An Abrupt Climatically Induced Rain Forest Fragmentation in West Africa during the Late Holocene', *The Holocene* 15(2): 190–199.

Schabel, H.G. 1990. 'Tanganyika Forestry under German Colonial Administration, 1891–1919', *Forest and Conservation History* 34(3): 130–141.

Schroth, G. and C.A. Harvey. 2007. 'Biodiversity Conservation in Cocoa Production Landscapes: An Overview', *Biodiversity and Conservation* 16(8): 2237–2244.

Scott, J.C. 1969. 'The Analysis of Corruption in Developing Nations', *Comparative Studies in Society and History* 11: 315–341.

———. 1998. *Seeing Like a State. How Certain Schemes to Improve the Human Condition Have Failed*. Yale University Press, New Haven.

Shanahan, T.M., J.T. Overpeck, C.W. Wheeler, J.W. Beck, J.S. Pigati, M.R. Talbot, C.A. Scholz, J. Peck and J.W. King. 2006. 'Paleoclimatic Variations in West Africa from a Record of Late Pleistocene and Holocene Lake Level Stands of Lake Bosumtwi, Ghana', *Palaeogeography, Palaeoclimatology, Palaeoecology* 242(3–4): 287–302.

Sharpe, B. 2005. 'Understanding Institutional Contexts to Define Research Questions. Settlement, Forestry, Identities and the Future in South-West Cameroon', in *Rural Resources and Local Livelihoods in Africa*, edited by K. Homewood. James Currey, Oxford.

Shaw, T. 1980. 'Prehistory', in *Groundwork of Nigerian History*, edited by O. Ikime. Heinemann Educational Books (Nigeria), Ibadan.

Shaw, T. and S.G.H. Daniels. 1984. 'Excavations at Iwo Eleru', *West African Journal of Archaeology* 14: 1–195.

Sheil, D., S. Jennings and P. Savill. 2000. 'Long-Term Permanent Plot Observations of Vegetation Dynamics in Budongo, a Ugandan Rain Forest', *Journal of Tropical Ecology* 16: 765–800.

Sheridan, M. and C. Nyamweru (eds). 2007. *African Sacred Groves: Ecological Dynamics and Social Change*. James Currey, Oxford.

Shore, C. and S. Wright. 1997. *Anthropology of Policy: Critical Perspectives on Governance and Power*. Routledge, London.

Siebert, U. and G. Elwert. 2002. 'Potentials and Recommendations for Combating Corruption and Illegal Logging in the Forest Sector of Benin, West Africa', paper presented at the Workshop on Illegal Logging in the Tropics, Yale University.

Sivaramakrishnan, K. 1999. *Modern Forests. Statemaking and Environmental Change in Colonial Eastern India*. Stanford University Press, Stanford, CA.

Smith, D.J. 2001. 'Kinship and Corruption in Contemporary Nigeria', *Ethnos* 66(3): 344–364.

———. 2007. *A Culture of Corruption: Everyday Deception and Popular Discontent in Nigeria*. Princeton University Press, Princeton.

Smith, W. 1744. *A New Voyage to Guinea*. J. Nourse, London.

Sowunmi, M.A. 1986. 'Change of Vegetation with Time', in *Plant Ecology in West Africa*, edited by G.W. Lawson, pp. 273–307. John Wiley & Sons, London.

———. 1999. 'The Significance of the Oil Palm (Elaeis Guineensis Jacq.) in the Late Holocene Environments of West and West Central Africa: A Further Consideration', *Vegetation History and Archaebotany* 8: 199–210.

Spichiger, R. and C. Blanc-Pamard. 1973. 'Recherches sur le contact forêt-savane en Côte D'ivoire: Étude du recru forestier sur des parcelles cultivées en lisiére d'un ilôt forestier dans le sud du pays baoulé', *Candollea* 28: 21–37.

Sprugel, D.G. 1991. 'Disturbance, Equilibrium, and Environmental Variability: What Is "Natural" Vegetation in a Changing Environment?', *Biological Conservation* 58: 1–18.

St. Barbe Baker, R. 1928. 'Silvicultural Experiments at Sapoba, Nigeria', *Empire Forestry Journal* 7(2): 203–208.

———. 1942. *Africa Drums*. Lindsay Drummond, London.

Stebbing, E.P. 1937. 'The Threat of the Sahara', *Journal of the Royal African Society* 36(145): 3–35.

Steege, H.T. 2003. *Long-Term Changes in Tropical Tree Diversity. Studies for the Guiana Shield, Africa, Borneo and Melanesia*. Tropenbos Series 22. Tropenbos International, Wageningen.

Steward, J. 1955. *Theory of Cultural Change: The Methodology of Multilinear Evolution*. University of Illinois Press, Urbana.

Steward, J.H., J.C. Steward and R.F. Murphy. 1977. *Evolution and Ecology: Essays on Social Transformation*. University of Illinois Press, Urbana and London.

Sunseri, T.R. 2009. *Wielding the Ax: State Forestry and Social Conflict in Tanzania, 1820-2000*. Ohio University Press, Athens, OH.

Swift, J. 1996. 'Desertification Narratives, Winners and Losers', in *The Lie of the Land: Challenging Received Wisdom on the African Environment*, edited by M. Leach and R. Mearns, pp. 73–90. James Currey, Oxford.

Szeftel, M. 1998. 'Misunderstanding African Politics: Corruption and the Governance Agenda', *Review of African Political Economy* 76: 221–240.

Talbot, M.R. and G. Delibrias. 1977. 'Holocene Variations in the Level of Lake Bosumtwi, Ghana', *Nature* 268: 722–724.

Thompson, A. 2005. *The Empire Strikes Back? The Impact of Imperialism on Britain from the Mid-Nineteenth Century*. Pearson Longman, Harlow.

Thompson, H.N. 1906. 'Forestry and Agriculture', in *Civil Service List and Handbook of the Colony and Protectorate of Southern Nigeria 1906*. Government of Southern Nigeria, Lagos.

———. 1910. 'Annual Report on the Forestry and Agricultural Departments of Southern Nigeria for the Year 1909', Forestry Department of Southern Nigeria.

———. 1911a. 'Annual Report on the Forestry Department for the Year 1910', Forestry Department of Southern Nigeria.

———. 1911b. 'The Forests of Southern Nigeria', *Journal of the African Society* 10(38): 121–145.

———. 1913. 'Annual Report of the Forest Administration of Southern Nigeria for the Year 1912', Forestry Department. Copies available from CO 657/1.

———. 1917. 'Annual Report on the Forest Administration of Nigeria for the Year 1916', Forest Administration of Nigeria.

———. 1919. 'Annual Report on the Forest Administration for the Year 1918', Forest Administration of Nigeria.

———. 1923. 'Annual Report on the Forest Administration of Nigeria for the Year 1922', Forest Administration of Nigeria.

———. 1925. 'Annual Report on the Forest Administration of Nigeria for the Year 1924', Forest Administration of Nigeria.

———. 1926. 'Annual Report on the Forest Administration of Nigeria for the Year 1925', Nigerian Forest Administration.

———. 1927. 'Annual Report on the Forest Administration of Nigeria for the Year 1926', Forest Administration of Nigeria.

Tignor, R.L. 1993. 'Political Corruption in Nigeria before Independence', *The Journal of Modern African Studies* 31(2): 175–202.

Tropp, J.A. 2006. *Natures of Colonial Change: Environmental Relations in the Making of the Transkei*. Ohio University Press, Athens, OH.

Turner, E.C. and W.A. Foster. 2009. 'The Impact of Forest Conversion to Oil Palm on Arthropod Abundance and Biomass in Sabah, Malaysia', *Journal of Tropical Ecology* 25(1): 23–30.

Udo, R.K. 1965. 'Sixty Years of Plantation Agriculture in Southern Nigeria: 1902–1962', *Economic Geography* 41: 356–68.

Unwin, A.H. 1920. *West African Forests and Forestry*. T. Fisher Unwin, London.

———. 2003. 'Forest Clearance in Early Colonial Benin and Ishan Divisions 1899–1919: Debunking the Shifting Cultivation Argument', paper presented at the University of Ibadan History Department.

———. 2005. 'Pre-Colonial Benin: A Political Economy Perspective', in *Pre-Colonial Nigeria: Essays in Honour of Toyin Falola*, edited by A. Ogundiran. Africa World Press, Trenton, NJ.

van den Boogaart, E. 1987. 'Books on Black Africa. The Dutch Publications and their Owners in the Seventeenth and Eighteenth Centuries', *Paideuma* 33: 115–126.

Vandergeest, P. and N.L. Peluso. 2006a. 'Empires of Forestry: Professional Forestry and State Power in Southeast Asia, Part 1', *Environment and History* 12(1): 31–64.

———. 2006b. 'Empires of Forestry: Professional Forestry and State Power in Southeast Asia, Part 2', *Environment and History* 12(4): 359–393.

van Gemerden, B.S., H. Olff, M. Parren and F. Bongers. 2003. 'The Pristine Rain Forest? Remnants of Historical Human Impacts on Current Tree Species Composition and Diversity', *Journal of Biogeography* 30: 1381–1390.

Vayda, A. and B. Walters. 1999. 'Against Political Ecology', *Human Ecology* 27: 167–179.

Verburg, R. and C. van Eijk-Bos. 2003. 'Effects of Selective Logging on Tree Diversity, Composition and Plant Functional Type Patterns in a Bornean Rain Forest', *Journal of Vegetation Science* 14(1): 99–110.
Vetch, C.F. 1912. 'Report on the Tapping of Communal Rubber Plantations for the Year 1911'. Copies available from No. 42 of Legislative Council Papers 1912.
Vincens, A., D. Schwartz, H. Elenga, I. Reynaud-Farrera, A. Alexandre, J. Bertaux, A. Mariotti, L. Martin, J.-D. Meunier, F. Nguetsop, M. Servant, S. Servant-Vildary and D. Wirrmann. 1999. 'Forest Response to Climate Changes in Atlantic Equatorial Africa during the Last 4000 Years BP and Inheritance on the Modern Landscape', *Journal of Biogeography* 26: 879–885.
von Hellermann, P. 2005. 'Things Fall Apart? A Political Ecology of 20th Century Forest Management in Edo State, Southern Nigeria', Ph.D. dissertation, University of Sussex.
———. 2010. 'The Chief, the Youth, and the Plantation: Communal Politics in Southern Nigeria', *Journal of Modern African Studies* 48(2): 259–283.
———. 2011. 'Reading Farm and Forest: Colonial Forest Science and Policy in Southern Nigeria', in *Engaging Colonial Knowledge: Reading European Archives in World History*, edited by R. Roque and K. Wagner, pp. 89–112. Palgrave Macmillan, Basingstoke.
von Hellermann, P. and U. Usuanlele. 2009. 'The Owner of the Land: The Benin Obas and Colonial Forest Reservation in the Benin Division, Southern Nigeria', *Journal of African History* 50(2): 223–246.
Walker, P.A. 2006. 'Political Ecology: Where Is the Policy?', *Progress in Human Geography* 30(3): 382–395.
Wardell, A. and C. Lund. 2006. 'Governing Access to Forests in Northern Ghana: Micro-Politics and the Rents of Non-Enforcement', *World Development* 34(11): 1887–1906.
Weir, A.H.W. 1938. 'Annual Report on the Forest Administration of Nigeria for the Year 1937', Forest Administration of Nigeria.
———. 1939. 'Annual Report of the Forest Administration of Nigeria for the Year 1938', Forest Administration of Nigeria. Copies available from Sessional Paper No. 34 of 1939.
Weiss, T.G. 2000. 'Governance, Good Governance and Global Governance: Conceptual and Actual Challenges', *Third World Quarterly* 21(5): 795–814.
White, L.J.T. and J.F. Oates. 1999. 'New Data on the History of the Plateau Forest of Okomu, Southern Nigeria: An Insight into How Human Disturbance Has Shaped the African Rainforest', *Global Ecology and Biogeography* 8(5): 355–361.
Wilks, I. 1993. *Forests of Gold: Essays on the Akan and the Kingdom of Asante*. Ohio University Press, Athens, OH.
Willis, K.J., L. Gillson and T.M. Brncic. 2004. 'Ecology: How "Virgin" Is Virgin Rainforest?', *Science* 304(5669): 402–403.

Wolf, E. 1972. 'Ownership and Political Ecology', *Anthropological Quarterly* 45(3): 201–205.
Wood, D. 1993. 'Forests to Fields: Restoring Tropical Lands to Agriculture', *Land Use Policy* 10(2): 91–107.
Worboys, M. 1979. 'Science and British Colonial Imperialism, 1895–1940', Ph.D. dissertation, University of Sussex.
———. 1996. 'British Colonial Science Policy, 1918–1939', in *Les Sciences Coloniales. Figures et Institutions*, edited by P. Petitjean. ORSTOM editions, Paris.
Worster, D. 1993. *The Wealth of Nature. Environmental History and the Ecological Imagination*. Oxford University Press, Oxford.
Yoon, C.K. 1993. 'Rainforests Seen as Shaped by Human Hand', *New York Times*, 27 July.
Young, C. 1988. 'The African Colonial State and Its Political Legacy', in *The Precarious Balance. State and Society in Africa*, edited by D. Rothchild and N. Chazan, pp. 25–66. Westview Press, Boulder, CO and London.
Zeven, A.C. 1967. 'The Semi-Wild Oil Palm and Its Industry in Africa', Agricultural Research Reports 689. Centre for Agricultural Publications and Documentation, Wageningen.
———. 1972. 'The Partial and Complete Domestication of the Oil Palm (Elaeis Guineensis)', *Economic Botany* 26: 274–279.

Index

Abacha, S. 67, 113
Abiola, M. 67
Aborigines Protection Society 49
afara, white and black, see *Terminalia superba*
Africa Forest Law Enforcement and Governance (AFLEG) 5
African Timber and Plywood Company (AT&P) 104, 111, 113–14
  See also United Africa Company (UAC)
Ainslie, J.R. 53, 98
Akpanigiakon 23–24, 30
Akpata, E.I.O 103
Akure 22
Albizzia spp 36
Alstonia boonei 36
Amowie, J. 68
Antiaris Africana 36
Asante Kingdom 20
Assamara 68, 70
avocado tree 36

Babangida, I. 14, 66
Benin City 7, 8, 22, 23, 25, 27, 30–31, 34, 41, 43, 51, 66, 67, 71, 73, 93, 112, 118, 119, 134
Benin Forest Scheme 54–57, 99, 101, 129
Benin Kingdom
  age grades 34
  bronzes 28, 43
  chieftaincy titles 23–25, 31–32
  early history 22–25
  European records of 26–29
  landscapes 22, 24, 26–30, 35–42
  land control 34
  late history 40–43
  medicine 39
  political structure 30–32
  slavery 31–32
  villages 34
  See also Edo language and myth of origin
Benin Native Administration 54, 60, 61, 104,
Benin Native Administration Forest Circle 99, 100
Biafra War 64, 130
biodiversity 14, 47, 60, 75, 76, 78, 80–82, 107, 108, 111, 121, 143, 145, 146
biomass 14, 76, 78, 80–82, 135, 143, 154
Bosumtwi, Lake 20
Brandis, D. 11
British Honduras 88
Burgess, H.B. 63
Burma 12, 41, 87, 126, 128, 130

Calabar 43, 49, 70
cameral sciences 11
Cameroon 15, 16, 23, 29, 147
carbon sequestering 14, 75, 76, 80
cassava 29, 33, 59, 151, 152
Ceiba Pentandra 35, 121–22
Celtis soyaruxii 36
Central African Republic 8
Chamberlain, N. 12
Chena plantations 128–29
Chrysophyllum africanum 36
Chukwogo, E.M.O. 103
civilian rule 14, 67, 68, 113
climate change 5, 14
cocoa 68–72, 74–76, 81, 118, 134, 137, 139, 142, 149, 157–58
coconut 37

Index

Collier, C.F. 63, 103–5, 109
Colonial Development and Welfare Act of 1940 55, 62, 103
community-based conservation 5, 17, 60–61, 143, 145, 147, 148–54, 158–60
Congo 55
Cote d'Ivoire 72, 88, 92
cotton 29, 128
crisis narratives 1, 3
Cross River National Park 147, 149
Cross River State 70, 73, 119
Crown land 48
cultural ecology 6

Dahomey Gap 20
Darling, P. 23, 30, 113, 153
decentralisation 5, 145
deforestation 1–2
depression 97
Deutsche Gesellschaft für Technische Zusammenarbeit (GTZ) 110
Diospyros spp 36
Dust Bowl experience 53
Dutch West India Company 27

Ebue Forest Reserve 57, 62
ecology 9
Edo language 7, 22
　　Edo myth of Origin 22
Egerton, W. 50
Ehor Forest Reserve 57, 62, 68
Ekenwan Forest Reserve 57
Ekiadolor Forest Reserve 57, 62
elephants 28, 155, 157
Entandrophragma cylindricum 36, 60, 101, 129
environmentality 12, 155
Etete 58, 64

Ficus asperifolia 36
fire 13, 35, 111, 113, 121
Ford Foundation 151, 152
Forest Ordinances
　　1901 Southern Nigeria Forestry Proclamation 48–49, 89
　　1908 Forest Ordinance 49, 89, 90, 93–94
　　1916 Forest Ordinance 52–53, 89, 90
　　1927 Forest Ordinance 54
Foucault, M. 11–12
France 11
Friends of the Earth 74, 160
frontier 82

Germany 11
Ghana 20, 29, 42
Gilli-Gilli Forest Reserve 53, 146
girth size 89–90, 115, 117, 119, 121
Glasgow Empire Exhibition 101
gmelina see *Gmelina arborea*
Gmelina arborea 110, 111, 113, 121, 130, 136
Gold Coast 88
　　See also Ghana
good governance agenda 2–3
governmentality 11–12
Guarea cedrata 36, 129
Gwatto 27, 41

hippopotami 28
historical ecology 9
Hitchens, P. 89
horses 28
human ecology 6

Ibadan 72, 97, 103, 110
Ibadan Forestry School 103
Ibo speaking people 70, 72
Ife 23
Ifon Game Reserve 119
Igbinedion, L. 67–68
Igbo-Iwoto Esie 22
Ighile, J.O. 67
Iguafole 33, 58, 67, 74, 139
Igueben Local Government Area 120
Igueze 33, 37, 38, 67, 74, 131, 138, 139
Iguobazuwa 33, 67, 69, 72, 116, 118, 119, 120, 131
Iguobazuwa Reserve 57, 62, 68, 74, 116, 134, 136, 139, 158, 160
Iguowan 33, 35, 58, 59, 69, 73, 74, 114, 116, 119, 120, 122, 135, 137, 139, 149, 151, 152, 153, 155, 157
Ijaw speaking people 33, 49, 62, 69, 149, 152

ikhimwin tree see *Newbouldia laevis*
Imperata spp. 137
India 11, 12, 13, 87, 97, 98, 102, 113, 155
International Labour Organisation (ILO) 74, 153
International Monetary Fund (IMF) 2
iroko tree see *Milicia excelsa*
iron 22
iron wood tree see *Lophira alata*
Irvingia gabonensis 36
Itsekeri 53
ivory 28
Iwo-Eleru 22
iya earthworks 23, 24, 30, 42
Iyayi Group 66–67, 68, 114–18, 152
Iyek'Ovia 20, 21, 33–34

Jamieson Reserve 49
   See also Sapoba

Kennedy, J.D. 97, 128–29
Khaya ivorensis 36, 60, 88, 101, 129
Khaya senegalensis 88
kola nut 36, 37, 134, 142
Korup National Park 147, 149

Lagos 28, 70, 72, 91, 99
Lagos Colony 50, 51, 54, 88, 89
Lagos Wood see *Khaya ivorensis*
land tenure 34, 56, 60–62, 65
Land Use Decree 65
Leventis Foundation 148, 156
Liverpool 48, 92
Lloyd, P. 149, 151–52
logging
   companies 88, 90, 100, 104–5, 106–7, 113–15
   effects on forest 95, 107–8
   illegal 106, 116–18
   regulations 89–90, 104–6, 115–18
   salvage felling 105, 108–9
Lophira alata 36
Lowe, R. 107

MacGregor. W.D. 99
mahogany see *Khaya ivorensis*
maize 29–30, 35, 128, 151

Malawi 17
Malaysia 55, 104
Mandoti 68, 69, 70
Mansonia 36
McNeil, D. 106–7
Michelin 66–67, 68, 74, 116, 152–53, 160
milhio 29
Milicia excelsa 35, 36, 39, 60, 92, 94, 101, 108, 129
military rule 14, 65, 66, 68, 113
Miller Brothers Ltd. 88, 90
millet 29–30
Ministry of Agriculture and Natural Resources (MANR) 68, 113, 118, 146
missionaries 61
Mohammed, General M.R. 65, 66
Mojo, J.I. 67, 68, 70, 74, 139, 156
Moor, Sir R. 11, 48, 50–51, 88–89
Mpiemu 8
Musanga smithii 36

Nancy 11
New Partnership for Africa's Development (NEPAD) 5
Newbouldia laevis 33, 37–38
Niger Coast Protectorate 7, 11, 12, 43, 48, 51
Niger Delta 22, 28, 74
Nigerian Conservation Foundation (NCF) 74, 148, 150–55, 157, 158
Nigerian Institute of Oil Palm Research (NIFOR) 78
Nikrowa 99, 100, 102, 148, 149, 151
Nyendael 27, 28, 29, 33, 40

oba
   Oba Akenzua II 54–56, 57, 61, 99, 100–1, 116, 129
   Oba Erediauwa 23
   Oba Esigie 24, 25–26, 33
   Oba Eweka I 23
   Oba Eweka II 52–53, 54
   Oba Ewuare 24–25
   Oba Ogolua 24
   Oba Ovoramwen 43, 52
   Oba Ozolua 24, 25
   Owner of the land 34, 52, 60

Index

Obahiagon, Chief 49
Obaretin Forest Reserve 53, 68
Obaseki, A. 93
Obaseki, G. 93, 100
obeche tree *see Triplochiton scleroxylon*
obobo tree *see Guarea cedrata*
Ogbomo, J.A. 67, 68, 70, 74, 139
ogiefa 33–34, 72
Ogiso period 23
Ohosu Forest Reserve 53, 78
oil (petroleum) 14, 64–65, 109
oil palm
   early evidence of 22, 24
   environmental impact of
      plantations 75–76, 78–79, 135, 154
   harvesting from wild trees 30, 35, 50
   plantations 50, 55, 61–62, 64–68. *see also* OOPC
   uses 35
Okomu Forest Project (OFP) 151–52
Okomu Forest Reserve 7
   creation 49
   extension 53, 57–59
   forest conversion 66–80
Okomu National Park 7, 154–58
Okomu Oil Palm Company (OOPC) 66–70, 72–74, 76, 78–80, 134, 150, 153–54, 156–58
Okomu Wildlife Sanctuary 148–54, 160
Oliha N'Udo 25, 64, 68, 71–72
Oliphant, J. N. 53–54, 99, 101, 102
Oluwa forest 36
Omo Forest Reserve 121
Omorodion, F. 149–51
Ondo Province 107
Ondo State 68, 69, 78, 114, 117–19, 121
Osanobua 22
Oshiomhole, A. 68
Osse River 49, 90
   See also Ovia River
Osse River Division 100, 102
Osse River Rubber Estates (ORREL) 64, 66, 68, 70, 72–73, 74, 76, 78–80, 116, 120, 134, 139, 152, 153–54, 156–57
ostrich 28
otien fruit tree *see Chrysophyllum africanum*

Overseas Development Agency 148
Ovia Cult 39, 58
Ovia North-East Local Government Area 65, 119
Ovia River 23, 33, 39, 49
   See also Osse River
Ovia South-West Local Government Area 7, 33, 116, 119, 131
Owan Forest Reserve 53
Oyo 39, 42

palm kernels 24, 35, 50, 92
palm oil 22, 29, 30, 50–51, 55, 61, 64–65, 66, 70, 73–74, 76, 78, 81, 92
   See also oil palm
palm wine 29, 35, 94
Permanent Crops Order 1937 61
Piedmont 111, 114
plantain 33, 35–36, 47, 69–72, 73, 75–77, 81, 118, 128–29, 139, 142–43, 147, 155, 157, 158
political ecology 2, 6, 8, 9
Portuguese 26, 27
Presco 68
Pro-Natura International 152
Protectorate of Southern Nigeria 12, 50, 87, 88, 91, 92
Punch, C. 29, 41

Redhead, J.F. 42
Reserve Settlement Officer 57–59
Richards, A. 63
Riconondendron africanum 36
Rosevear, D. R. 63, 103–4, 106–9
Ross, R. 99–100
Royal Niger Company 12
royalties 54, 89, 90, 93, 95, 140
rubber
   environmental effects of plantation 75–76, 78–80, 135, 137–38, 154
   harvesting of wild rubber 48, 51, 88, 89
   plantations 51, 55, 61–62, 64, 66, 67, 68, 93, 94–95, 134, 153, *see also* ORREL
   prices 55, 80, 158

sacred groves 37, 39, 58, 143
salvage felling *see* logging
sapele wood *see Entandrophragma cylindricum*
Sapoba Forest Reserve 97, 129
  See also Jamieson
Sapoba Research Station 97–98, 99, 100, 104, 110, 128–29
sawmills 101, 106, 111, 114, 117, 119, 121
Schlich, W. 11
Scott, J. 12, 17
Shaw, T. 22
shifting cultivation 36, 48, 53, 127–28, 132, 140
Sierra Leone 16, 42
silk cotton tree *see Ceiba pentandra*
slavery *see* Benin Kingdom
small pox 40
Sobo *see* Urhobo
Socfinco 66
soil 13, 35, 75–76, 78, 145, 148, 150, 154
Somerville, G.M. 104
South Africa 13
South Asia 13, 111
Southeast Asia 12, 17
spear grass *see Imperata spp.*
St. Barbe Baker, R. 54, 96–98, 104, 128
Stebbing, E.P. 53
Stockholm 98
Structural Adjustment Programmes 2, 3, 14, 15, 70, 134
Swedish International Development Cooperation Agency (SIDA) 110
sweet potato 29
Sykes, R.A. 99

Tanzania 5, 146
teak *see Tectonia grandis*
Tectonia grandis 95, 110–11, 113, 121, 130, 131, 136
Terminalia superba 36, 60, 120
Third National Development Plan 65
Thompson, H. N. 12, 41–42, 48–52, 60, 87, 90, 92, 94–97, 99, 101, 104, 108, 111, 127–28
timber boom 105–6, 109
Triplochiton scleroxylon 36, 60, 101, 104, 114

Tropical Forest Action Plan (TFAP) 113
Tropical Shelterwood System (TSS) 104, 106–8, 109–10, 130

Ubinu 7
Udo Forest Reserve 120
Udo Town 7
  foundation and early history 23–25
  recent development 71–74
Ugie festival 25
umbrella tree *see Musanga smithii*
United Africa Company (UAC) 100, 101, 102, 104, 106–7, 111, 113
  See also African Timber and Plywood Company (AT&P)
United Nations 5, 110, 113
  Food and Agriculture Organisation (FAO) 110
  Programme on Reducing Emissions from Deforestation and Forest Degradation in Developing Countries (REDD) 5
Urezen 24, 33, 39, 58
Urhobo 50, 53
Usonigbe Forest Reserve 57, 62
uzama 23, 24, 32, 71

Warri Kingdom 28
watershed protection 12, 13, 145
Weir, A.H.W. 57
Western Metal Products Company Ltd (WEMPCO) 114, 119, 121
working plans 12, 47–48, 96, 98–100, 102, 104–5, 106–7, 111–12, 113, 115, 141
World Bank 2, 110
World Rainforest Movement 74, 160
World War One 51, 52, 92, 96
World War Two 61, 62, 101, 102
World Wildlife Fund 148, 149

yam 22, 29, 30, 35–36, 59, 69, 70, 128, 133–34, 142
Yoruba
  cocoa farmers 47, 68–72, 74, 76, 81, 137, 149, 156
  kingship 23
  towns 20, 31, 42